Y0-BVN-142

INTERNATIONALIZATION

List of Faculty Members

Ulf Lantzke (Chairman), Executive Director of the International Energy Agency, Paris

Harold M. Agnew, Director of the Los Alamos Scientific Laboratory, University of California

Ross Campbell, Chairman of Atomic Energy of Canada, Ltd.

D. A. V. Fischer, Assistant Director General for External Relations International Atomic Energy Agency, Vienna

René Foch, Former Director of Euratom, Brussels

Russell W. Fox, Ambassador at Large for Australia for Nuclear Nonproliferation and Safeguards

Bertrand Goldschmidt, Advisor to the French Commissariat à l'Energie Atomique, Paris

Gunter Hildenbrand, Head of Nuclear Fuel and Nuclear Fuel Cycle Division, Kraftwerk Union A.G., Erlangen, Germany

Joseph S. Nye, Jr., Deputy to the U.S. Undersecretary of State for Security, Washington

Dwight J. Porter, Director of International Government Affairs for Westinghouse, Washington

Terence Price, Secretary General of the Uranium Institute, London

J. Robert Schaetzel, Former U.S. Ambassador to the European Community

Glenn Seaborg, Former Chairman of the U.S. Atomic Energy Commission and recipient of the Nobel Prize in Chemistry.

Internationalization

An Alternative to Nuclear Proliferation?

edited by
Eberhard Meller
International Energy Agency

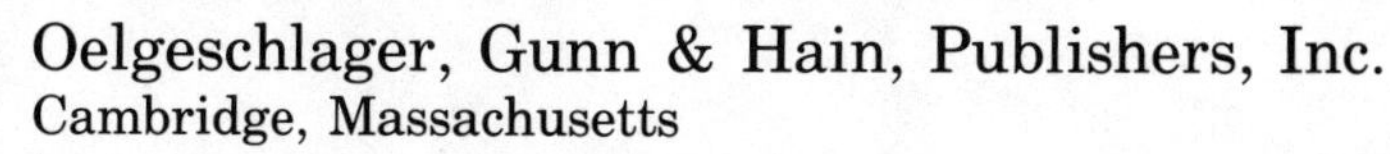

Oelgeschlager, Gunn & Hain, Publishers, Inc.
Cambridge, Massachusetts

International Standard Book Number: 0-89946-049-6

Library of Congress Catalog Card Number: 80-17265

Printed in the United States of America

Library of Congress Cataloging in Publication Data

Salzburg Seminar on American Studies, 1978.
Internationalization, an alternative to nuclear proliferation?

1. Nuclear nonproliferation—Congresses. 2. Atomic power—International control—Congresses. I. Meller, Eberhard. II. Title.
JX1974.73.S24 1978 327.1′74 80-17265
ISBN 0-89946-049-6

The views expressed are those of the editor or faculty members and do not necessarily represent the position of the International Energy Agency.

Contents

List of Figures

List of Tables

Foreword

The name "Salzburg Seminar in American Studies" accurately described the first five-week session held at the Schloss Leopoldskron in the summer of 1947. All of the faculty was American (and all from Harvard, for that matter), whereas the fellows were overwhelmingly European. Through discussions of the humanities, the social sciences, and law, with stress on American thought and developments during the war years, a start was made in reestablishing private American and European relations, which had been severed by more than five years of war.

As the Seminar's role grew, its emphasis became directed more and more along the lines of American and European studies. More Europeans joined the faculty, and more fellows came from Eastern Europe, the Far East, the developing countries, and—starting in 1979—the Middle East.

While law and the humanities have continued to constitute a firm part of the curriculum, the sessions increasingly have included topical subjects that concern both the highly industrialized and the developing Third World countries. Thus, it was natural that in 1977 the decision was made to hold a session titled "Is Internationalization the Alternative to Nuclear Proliferation?" The distinguished international faculty, chaired by Dr. Ulf Lantzke, Executive Director of the International Energy Agency, is listed on page ii of this volume. The fellows from twenty-five countries, mentioned by Eberhard Meller in his preface, are an equally talented group; however, they are young professionals just starting their careers, and hence for the most part their names are not as yet very well known.

Dr. Lantzke, in his concluding statement (pages 33–38 of this volume) answers the question in the title with "yes—but a yes well understood". He and the other faculty members have made an outstanding contribution to the efforts of all of us to understand this complex and all-important issue.

Dr. Lantzke states, "Nuclear power's contribution to energy supply is indispensable." He goes on to add, "The public has not been well informed and in many cases misinformed. Public education is a difficult task and we cannot bring everybody for two weeks to Schloss Leopoldskron." He sums up this point by saying, "The confidence of the public in institutions . . . is a precondition to greater public acceptance of nuclear [energy]." So, in addition to all the complex issues raised in this volume, we cannot avoid the fact that no rational "solutions" are possible if the public lacks confidence in the national and international institutions. And it is clear that the existing institutions must be revitalized and made more meaningful if we are to make progress regarding this and other complex issues.

The humanities, the social sciences, and law all have an essential part to play as we recognize that the proliferation of nuclear weapons and other urgent problems can be adequately tackled only in societies in which the population understands and plays a vital role in the entire sweep of issues that confront an interdependent world. The Salzburg Seminar intends to continue to hold sessions on pressing topical subjects, while recognizing the interrelationship of these subjects with the humanities and other disciplines that can help create an alert and informed—and international—body politic. "Internationalization," together with institutions with adequate charters of responsibilities that are staffed by courageous and imaginative people, is essential in our search for solutions to nuclear and other complex issues that extend far beyond country boundaries.

The Seminar will continue to hold at least one or two sessions each year on the humanities (music, theater, literature, philosophy, poetry, for example) in the conviction that, in fact, "man is the measure of all things" and that the humanities will help us understand ourselves. But we will continue to include contemporary history and government and topical subjects, such as energy, in our curriculum. For example, from April 21 to May 1, 1981, a session on "Energy and Global Security" will be held at the Schloss Leopoldskron, with Dr. Lantzke again serving as a key faculty member. In an effort to better comprehend—and, we hope, improve—the political process in our countries, there will be a session on "The Process of Selecting Our Political Leaders," to be held August 9–28, 1981, with Professors Daniel Bell and Karl Deutsch included in the faculty. And in 1982 a session will be held on "The Role of International Organizations and National Governments in Foreign Policy Decision Making." The issue of the proliferation of nuclear weapons and the related issue of the role of nuclear (nonmilitary) energy are bound to be included in future sessions. Such sessions will certainly benefit from this volume, which sets forth the views of well-informed faculty and fellows who examined this difficult problem in the spirit of free inquiry during a two-week session at the Schloss Leopoldskron.

John Wills Tuthill
President
Salzburg Seminar in American Studies

Preface

For three decades the Salzburg Seminar on American Studies has provided a discussion forum for improving international understanding and cooperation among nations. In 1977, when the Seminar first planned the 1978 conference—Is Internationalization the Alternative to Nuclear Proliferation?—world security and disarmament were to be the primary focus. However, in 1977 the debate on nonproliferation of nuclear weapons became more intense and often heated, particularly in the United States. Moreover, the differences between the United States and its European allies reached a critical stage. These differences not only increased the mistrust between allies but also threatened to hamper the development of nuclear programs of the Western world.

A common understanding between all interested nations of their future nuclear programs was urgently needed. It became clear that better information and facts and continued exchange of views and perceptions were necessary to improve understanding at all levels. While the official debate at the Vienna International Fuel Cycle Evaluation (INFCE), which was proposed by President Carter, had been organized as a technical and analytical study, the Salzburg Seminar provided a forum where more emphasis could be placed on the social and political aspects of nuclear energy.

The two-week conference held in the autumn of 1978 proved to be one of the most important informal meetings of the last few years. The varied backgrounds and expertise of the participants, coupled with the open atmosphere of the conference, resulted in in-depth discussions on diverse topics of great interest and value. The seminar was a microcosm of the wider debate. On a smaller scale, established national attitudes were evident as well as those of the nuclear industry, environmentalists, and international organizations. The seminar included some forty-five young professionals from twenty-five Eastern and Western countries in Europe, as well as from Australia, Iran, Canada, and Latin America, and a faculty of leading personalities in the field of nuclear energy worldwide.

A primary focus of the seminar was the assessment of possible improvements in present world nonproliferation policies. In particular, the notion of internationalization of sensitive parts of the nuclear fuel cycle was analyzed in detail. Following a careful review of past and current nonproliferation policies, the seminar examined the current controversy between the United States and its European allies with reference to the development of fast breeder reactors and plutonium recycling. Particular emphasis was given to how the U.S. Non-Proliferation Act of 1978 affected U.S.–Euratom relations. Industrial governmental relationships were also analyzed, as well as safety and the special needs of developing countries.

Special emphasis was placed on the problems of international corporations and, in particular, on the need for a new climate of trust and stability in international nuclear relations. Seminar participants reached a number of important conclusions on corporate problems that are presented in the Seminar Reports, prepared as a consensus paper by the members of the various groups. These reports summarize the discussions of the three seminar groups, which were held in addition to lectures given by the faculty members. A fourth seminar group, which was devoted to the history of nonproliferation and the lessons to be learnt by past history, did not issue a report.

The complete text of the lectures given by the faculty members are presented as Part II of the book and cover the whole spectrum of the nonproliferation debate, ranging from the role of nuclear power in energy supply to the complexity of nuclear trade and the impacts of the spread of nuclear weapons. The lecture material, arguments, and issues raised remain valid today because the problem has not yet been solved. The background and basis of current discussions can be found in the analysis and conclusions of these papers. Further, the Introduction and Overview not only introduces the subject matter but also reviews developments in the international nonproliferation debate, particularly the recent results of INFCE, since the seminar was held.

Eberhard Meller
Paris
April 1980

INTERNATIONALIZATION

Part I

Introduction and Overview

*Eberhard Meller**

Nonproliferation of nuclear weapons is still one of the most controversial and complex international issues of our times. This has become even more evident as a result of the events in Iran, which once again emphasized the gravity of the energy problem, encouraging further expansion of nuclear power. The events in Afghanistan also have revealed the fragility of global balances, emphasizing the insecurity and instability of different regions. Nonproliferation is commonly defined as prevention of the further spread of nuclear weapons. It includes so many conflicting elements—such as energy, military security, market forces, and governmental interference, as well as national aims and international commitments—that there are no easy solutions to the problem. This becomes evident when we look at the history of the development of nuclear power.

HISTORY, OR LESSONS TO BE LEARNED

The nonproliferation problem has been an aspect of international cooperation in the field of nuclear energy since the end of the

*Principal Administrator, International Energy Agency, Paris

Second World War. However, the problem has changed in character and temperament over the years. It has considerably influenced the spirit and the atmosphere of international cooperation, not only in the nuclear field but also in the overall relationships among nations. Periods of secrecy and denial have alternated with periods of openness and cooperation.

It is quite understandable that after the terrible experience of the Hiroshima and Nagasaki atomic bombs, the initial period of dealing with world nuclear policy would be characterized by secrecy and the maintenance of a monopoly. The technology holders, the United States and United Kingdom, tried to control this dangerous technology with a policy of denial of information about and transfer of nuclear material, equipment, and technology, while attempting as well to internationalize substantive parts of the nuclear fuel cycle. However, the Acheson-Lilienthal plan, a U.S. proposal that suggested an internationalization of all nuclear activities worldwide, failed since the Soviet Union refused to accept the restriction on national sovereignty that was implied in such a plan. Thus humanity lost its last chance to live in a world free from atomic weapons. As a footnote to this history, it should be noted that the exclusion of Canada from the U.S.–U.K. agreement on the protection of nuclear information led to the Canadians' decision to develop their own reactor system, the Candu reactor.

The second phase commenced in 1954 when President Eisenhower announced his Atoms for Peace program. This period, lasting about twenty years, could be called the period of confidence and openness and was certainly called by some observers the golden years of nuclear cooperation. But the world situation had changed. In addition to the United States, the Soviet Union, Great Britain, and France became members of the nuclear weapons club, and a further spread of nuclear knowledge and technology was unavoidable. Consequently, the Atoms for Peace program took a different approach. The revised goal was to open up a new energy source under conditions in which national and international interests were safeguarded against unlawful diversion and clandestine use of fissile material to produce nuclear explosives. The United States offered assistance for peaceful utilization of nuclear material on the condition that this assistance was not channeled toward military purposes.

At the center of this verification system was a set of safeguards, including on-site inspection of peaceful nuclear activities by inspectors from outside the host country. This was one of the reasons for creating in 1957 the International Atomic Energy Agency (IAEA) in Vienna, which was assigned the task of both promoting international cooperation in the peaceful utilization of nuclear energy and implementing the safeguard system. In replacing the bilateral safeguards arrangement that was relied on initially, the agency, which today has 125 members, became a cornerstone of the nonproliferation policy. The technical objective of these safeguards is "the timely detection of the diversion of significant quantities of nuclear material from peaceful nuclear activities to the manufacture of nuclear weapons or of other nuclear explosive devices or for other purposes unknown, and deterrence of such diversion by early detection."

The second major nonproliferation tool established by the international nuclear community was the Non-Proliferation Treaty (NPT), negotiated 1965–1968 and put into force in 1970. The NPT formally forbids building of nuclear weapons or peaceful nuclear explosives by nonnuclear weapon states and requires them to accept mandatory international safeguards on all their peaceful nuclear facilities. In order to encourage nonnuclear weapon states to ratify the NPT, it stated that any state accepting the treaty conditions would benefit from commitments that its peaceful nuclear development would not be impeded and that it would have the right to take part in "the fullest possible exchange of equipment, materials and scientific and technical information on the peaceful uses of nuclear energy."[1]

The number of NPT member states steadily increased. In addition to the three nuclear weapon states (United States, United Kingdom, and the Soviet Union), 107 nonnuclear weapon states have signed the NPT to date, so that it now includes the membership of the most advanced nonnuclear weapon states. The only countries with existing or planned nuclear programs that have not signed the NPT are Argentina, Brazil, Spain, India, South Africa, and Israel. There are only five states in the world besides the nuclear weapon states that have significant nuclear activities and that are not subject to IAEA safeguards—namely, Egypt, India, Israel, South Africa, and Spain.[2] Although China and France are not members, France has declared that it will act as if it were a party of the NPT.

During this period of trust and confidence, no parts of the fuel cycle, including the sensitive steps of enrichment and reprocessing, were prohibited. It was generally accepted, even by the United States, that the separation of plutonium from spent fuel and its use as a fuel in conventional reactors (recycling) was a natural component of an efficient light water reactor cycle, as well as a necessary step in effective nuclear waste disposal and a logical move toward the anticipated development of fast breeder reactors fueled by plutonium. The resulting climate first enabled many countries to accept the renunciation and discrimination between nuclear weapon and nonnuclear weapon states (haves and have nots) inherent in the NPT; second, it generated the free flow of all technique; and third, it supported the development of nuclear power on a grand scale in more and more countries.

Confidence in this approach eroded in the mid-1970s. The demise was triggered by the Indian nuclear explosion in 1974, which could be regarded as a "by-product" of Canadian-Indian cooperation. Moreover, new competitors in the field of nuclear technology appeared on the market, breaking the U.S. monopoly. Germany and France began to export, or at least to negotiate the exportation, not only of nuclear reactors but also of pilot enrichment and reprocessing facilities—for example, with Brazil, Iran, Pakistan, and South Korea. Doubts arose, especially in the United States and Canada, as to whether safeguards would still be adequate to deal with plutonium or the spread of enrichment facilities. This perception was based on a fear that existing safeguards could not provide sufficient detection methods to stop the acquisition of nuclear weapons. Even if detection were possible, the warning would not come in time for an international reaction before the realization of a weapons capability. It was argued that although there are simpler and more expeditious routes for acquiring nuclear weapons, the existence of sensitive commercial nuclear facilities and materials within national borders would increase the potential for building weapons, thus giving a nation the option of embarking upon a weapons program if it decided to do so. Moreover, countries would have to assign a higher probability to the possession of weapons by potential adversaries, thereby destabilizing international relations.

Unfortunately, this new perception, which could logically call for a stop or limitation of the world use of nuclear power, coincided with the increased concern of nations about their energy needs and supplies after the 1973 OPEC oil embargo. Ironically, this concern also led to fantastic nuclear power forecasts (e.g., eight reactors for Bangladesh), which in turn increased the fear that existing institutions would no longer be able to handle such a situation. Attempts were made to reach a common understanding among the nuclear exporting countries. The "London Guidelines" were set up in 1975, but only published in 1978 by the so-called London Supplier Club (an informal group of fifteen nuclear exporting states that meet regularly to coordinate their exports and their safeguard policies in support of the NPT objectives and provisions). They advocated a set of minimum guidelines for the export of nuclear material and technology. However, in the United States and Canada the common opinion was that IAEA safeguards and the NPT were inadequate and that therefore a technological barrier to proliferation was needed. This new approach culminated in the temporary Canadian embargo of natural uranium to Euratom and Japan in 1977 and the enactment of the Non-Proliferation Act in 1978 by the U.S. Congress. Both countries aim to renegotiate existing agreements so as to have a say in the transfer of nuclear material and technology by the recipient country (e.g., prior approval right to reprocessing).

In order to back this new policy, the United States had already decided in 1977 on the indefinite deferral of commercial reprocessing and plutonium recycling as well as on a slowdown and redirection of breeder development away from the plutonium-fueled cycle. The newly elected Carter administration made the review of nonproliferation policies one of its priority concerns. To counter foreign criticism, the United States offered verbal assurances that they would again fulfill their role as a reliable supplier of nuclear fuel. The criticism outside the United States, not only in Europe and Japan but also in the developing countries, emphasized that the new approach was based solely on the specific U.S. situation with its vast energy resources and did not take into account the different energy circumstances of other countries. This new policy was also sometimes regarded not only as antiplutonium but also as antinuclear, which therefore threatened the nuclear programs of countries already in difficulties because of public opposition. As a consequence, this new U.S. policy would endanger the one essential pillar of the overall energy supply situation. This, of course, also affected other areas of international cooperation.

Aware of this dangerous conflicting situation, the seven leading countries of the Western world (the United States, the United Kingdom, Canada, Germany, Italy, Japan, and France) agreed at the London Economic Summit in 1977 on the proposal by President Carter to undertake a major international fuel cycle evaluation (INFCE) in which all countries could participate and that would seek to reconcile future nuclear energy needs with nonproliferation concerns. The aim was to gain a breathing space and to cool down the often heated discussions by investigating ways to make present fuel cycles more resistant to proliferation or to find new ways, either technical or institutional, that would not hamper further expansion of the peaceful use of nuclear energy. The two-year study, in which more than 500 experts from sixty nations and five international organizations took part and which issued eight working group reports, a summary, and an overview, ended in February 1980. This conference turned out to have been an unprecedented international undertaking in both its scope and its objective. For the first time in the history of international cooperation, a highly disputed and complex international issue such as nuclear power had undergone a thorough international analysis that included the assessment of nuclear power projections, uranium requirements, uranium reserves, other nuclear fuels and production capacities and a comparison of various nuclear fuel cycles in terms of their potential proliferation hazards and the processing techniques employed in sensitive areas. Although the study was not a negotiation, but a technical and analytical evaluation, INFCE has not only dispelled many of the tensions that disturbed the international climate in 1977, but has also clarified many points and offers governments a collective analysis—all reports being consensus reports—of possible approaches and possible solutions, particularly in the institutional field.

The Salzburg Seminar met at the half-time of the INFCE deliberations in Vienna. The observations and conclusions reached at the Salzburg Seminar are similar to those reached by INFCE. This similarity shows that the Salzburg Seminar really was a microcosm of the wider debate on nuclear nonproliferation. The following comparison of the results of the two-week Salzburg Seminar with those of the two-year INFCE conference, demonstrates that the points made in Salzburg remain valid and critical to any discussion of nuclear proliferation.

CONCLUSIONS OF THE SALZBURG SEMINAR AND OF INFCE

Basic Conclusions

An overwhelming majority of both the Salzburg Seminar participants and all the countries and international organizations participating in INFCE agreed that the contribution of nuclear power to energy supply is indispensable, given the overall world energy supply and demand situation and the future prospects. The communique of the Final Plenary Conference of INFCE stated that "nuclear energy is expected to increase its role in meeting the world's energy need and can and should be widely available to that end."[3]

This statement can be interpreted as a new unified international commitment to nuclear power. It comes at a time when the contribution of nuclear power to the mix of world energy supplies is essential to overcome oil shortages expected in this century. No other technology is ready for a large-scale use. Today the expansion of nuclear energy has run into difficulties in many countries, especially after the Three Mile Island incident. INFCE nuclear power projections, based on 1977–1978 official national forecasts (adjusted only for 1985 and 1990 for member countries of the International Energy Agency), range from 850 GW to 1200 GW installed capacity in the year 2000 within the world outside centrally planned economic areas (WOCA). The low INFCE projections would represent more than two-thirds of current OPEC oil production and about 14 percent of the projected total energy requirements of WOCA. A nuclear depression and the slippage of nuclear programs have gripped most industrialized countries in the last few years. Rising costs, reduced demand for electricity, opposition by environmentalists, and concerns for safety have contributed to this depression. Even the low nuclear power projections by INFCE, which are essential to meet basic energy supply requirements, are not achieveable unless major and drastic improvements in the costs, siting, lead times, and acceptability of nuclear energy occur immediately. Concerning nonproliferation, two conclusions should be drawn. First, with critical shortages of oil anticipated later in this century, nonproliferation objectives can hardly be achieved by limiting world use of nuclear power. Second, slower development of nuclear power in the world has lessened the urgency for construction of additional fuel cycle facilities. The world has gained some time to implement an improved and internationally agreed upon nonproliferation regime.[4]

The Slazburg Seminar agreed that there are no "technical fixes" to prevent nuclear nonproliferation. The only thing that can be done is to make it technically more difficult for nations to divert nuclear materials from peaceful uses toward nuclear weapons. Technical improvements in the fuel cycle would only stop terrorists and would certainly not deter governments who are absolutely determined to construct nuclear explosive devices, particularly since the reliability of such a device does not need to be proved by testing.

INFCE also did not come up with "technical fixes," although this was one of the main goals of the two-year study. Some fuel cycle modifications were suggested. They included making plutonium less accessible (e.g., through colocation—the location of different nuclear fuel cycle facilities on the same site—and coprocessing—a modification of reprocessing in which plutonium does not exist in separated form). But such modifications would be effective primarily in dealing with theft, not in making it more difficult for nations to get weapons-usable materials.

Regarding risks, INFCE also established that there are proliferation risks in each fuel cycle and that it is not possible to reach a general judgment as to which particular fuel cycle entails more risk of proliferation than another. It will therefore not be possible to restrict on nonproliferation grounds the use of nuclear energy to one specific fuel cycle—for example, to the light water reactor, once through cycle. Moreover, different national situations would require different choices, so that the need and the timing of breeder development would vary among countries depending on their technical infrastructure, electric grid size, confidence of access to uranium resources, and other factors.

This leads to the common conclusion of both the Salzburg Seminar and INFCE that proliferation is basically a political matter and that if a nation elects to develop nuclear weapons, it could pursue this development without misusing civilian nuclear power facilities. More emphasis has therefore to be laid on substitution incentives for nations wishing to embark upon a nuclear weapons program. In contrast to INFCE, which could not address such a sensitive issue, the Salzburg Seminar identified the following major incentives that have to be countered by international policies (e.g., arms control and disarmament and "umbrella" policies):

1. Survival—in this context, countries such as Israel, South Africa, South Korea, Taiwan, and Pakistan were mentioned.
2. Prestige.
3. Regional leadership, the quest for military strength ("lust for power").
4. Security—to acquire additional bargaining power in exchange for stopping the political or military motives of a potential adversary.

Finally, both the Salzburg Seminar and INFCE concluded that a new climate of trust and stability must be restored in order to bring about fruitful cooperation. This was regarded as particularly important to nuclear trade, since concerns over security of supply have a direct impact on decisions of world nations concerning the whole nonproliferation problem.

Possible Improvement of the Existing Nonproliferation Regime

One of the basic premises of the seminar was that a radical new approach would not solve the nonproliferation problem. Improvements of the present regime would have to be built upon current institutions such as IAEA and safeguards such as NPT, London Suppliers Club, and other bilateral agreements.

This cautious attitude is also reflected in the INFCE reports. In particular, the safeguard system, based on either bilateral or, preferably, multilateral agreements, was regarded as a central feature of current and future nonproliferation regimes. In general, INFCE found the present safeguards system satisfactory, although developments and improvements were foreseen in technologies for uranium enrichment, industrial-scale reprocessing, and fuel fabrication for light water and breeder reactors.

Other possible improvements were also identified, by both the seminar and INFCE. These include assurance of supply, common nonproliferation undertakings, and institutional arrangements such as multinationalization of parts of the fuel cycle and uranium back-up arrangements. The goal of all measures would be to restore the stability and confidence in international cooperation. For, as Mr. Justice Fox points out (Chapter 8), the fear that a country may proliferate or could be in a position to develop nuclear weapons is probably a more important aspect of the problem than the actual risk of the spread of nuclear weapons.

Assurance of Supply

Assurance of supply and assurance of nonproliferation are complementary. Not only do effective nonproliferation assurances facilitate supply assurances, but the nonproliferation commitments of any country may be considered stronger when that country must rely on international markets for a part of its nuclear fuel supply. Moreover, greater assurance of supply can also contribute to nonproliferation objectives by reducing the pressures for a worldwide spread of enrichment and reprocessing facilities. Lack of assurance of long-term supply has in some cases already motivated countries to adopt policies of fuel cycle self-sufficiency earlier than would be required by their optimum economic and technical development schedule.

INFCE also noted that for most countries, concern about assurance of supply was not the result of commercial defaults or market failure but the result of government intervention in pursuit of national policies and objectives. This government intervention was usually associated with nonproliferation goals, but sometimes also with other defined national policies.

A recognized fact of life is that governments are not likely to give up the possibility of intervening in supply arrangements when they perceive it to be in their national or international interest to do so. Therefore, the main task of the years ahead will be to reconcile these different attitudes and perceptions. In accordance with both the ideas and the words expressed at the seminar, INFCE concluded that such common approaches, which could be expressed initially through practices of states and bilateral agreements, might eventually become joint declarations, codes of practice, or other multilateral or international instruments. This could lead to more formal measures, directed at ensuring secure access to nuclear materials, services, equipment, and technology under internationally accepted, effective nonproliferation conditions. Such an evolutionary process—building on existing instruments, institutions, standards, and practice—might lead in a practicable and conducive way toward a more certain nonproliferation regime in which national export and import policies might be implemented in a manner acceptable to both supplier and consumer countries.

While the seminar could not go into more detail about such common approaches during its two-week discussion, INFCE established nine fundamental criteria—or "matters" as they are called in the final summary and overview report. These nine criteria could form the basis of commonly agreed nonproliferation undertakings, particularly since most of them are also accepted within the framework of the NPT and IAEA. They include:

(a) Undertakings on the peaceful uses of nuclear materials, equipment and technology and verification of these.
(b) Undertakings not to develop or acquire nuclear weapons or nuclear explosive devices.
(c) Undertakings not to acquire, manufacture or store nuclear weapons or to help any country to do so.
(d) Undertakings with respect to the application of IAEA safeguards, including the requirements for nuclear materials accountancy and control and the implementation of any eventual IAEA system for storage of excess plutonium.
(e) Adequate levels of physical protection.
(f) Conditions governing the establishment and operation of certain stages of the nuclear fuel cycle and the management of their associated materials, including those stages based on international or multinational institutions or on national enterprises that fulfil a set of internationally or multilaterally agreed obligations.
(g) Duration of non-proliferation undertakings and controls.
(h) Sanctions and other measures to be applied in the case of a breach of non-proliferation arrangements.
(i) Undertakings regarding transfer and retransfer of supplied materials, equipment and technology, and their multilabelling and safeguards contamination implications.[5]

However, the 1978 enactment of the U.S. Non-Proliferation Act, which unilaterally changed the conditions in existing bilateral agreements, showed that such nonproliferation undertakings are subject to changes. This U.S. policy was criticized both at the seminar and in a more diplomatic way at INFCE. In the light of this experience, however, it was considered desirable for governments to develop mechanisms to manage changes in nonproliferation policies. The mechanism should be designed to reduce to a minimum the risk that if such changes should lead to disagreement between supplier and consumer countries, these changes would not also lead to interference with suppliers.

A number of possible mechanisms have been suggested for updating nonproliferation undertakings and conditions when necessary:

(a) Provision in intergovernmental agreements for, or a joint declaration of intent to conduct informal consultations among the parties to determine if changes are necessary, on the basis of which specific amendments might be contemplated.
(b) Provision in intergovernmental agreements for periodic review by the parties involved, possibly followed, if necessary, by amendments of non-proliferation undertakings and conditions in such agreements.
(c) Provision in intergovernmental agreements for the adoption of non-proliferation undertakings and conditions agreed on by multilateral review, to the extent that all governments party to the agreements have subscribed to them.
(d) The inclusion in intergovernmental agreements of contingency provisions under which further non-proliferation requirements would be introduced or existing requirements modified in response to particular developments.[6]

Assurances of supply could be enhanced if the adoption of such mechanisms were to be complemented by guarantees regarding continuity of supply during the renegotiation process. Suggestions that were discussed included:

(a) Undertakings by the parties to an agreement not to refuse export or import licenses under the terms of established contracts if the other party guarantees to accept amendments to non-proliferation conditions identified from time to time, in accordance with the mechanism agreed by the parties either bilaterally or within a broader international framework.
(b) Undertakings by the parties that any proposal for the extension or the amendment of non-proliferation requirements would not affect the issue of export and import approvals before the amendment mechanism has led to a consensus of the parties to the agreement.
(c) Undertakings that the parties will not interfere with deliveries under existing contracts for some reasonable period following a proposal for the extension or amendment of non-proliferation conditions, for example until it was clear that negotiations had reached an impasse.[7]

The special problem of the right of prior consent was also raised at INFCE and was recognized as having a negative impact upon the assurance of fuel supply and a consequent adverse effect upon national nuclear programs if exercised arbitrarily. Such a right, which is part of the U.S. and Canadian nonproliferation policy, gives the supplier country a veto over the retransfer of nuclear fuel to third countries and/or reprocessing of fuel supplied by them to consumer countries. It is not clear whether such a right would and could in fact be exercised like a "rubber guillotine," as described at the seminar by Joseph Nye (see Chapter 5), if there were no criteria. According to him the guillotine would come down, but no real harm need be done if the safeguards records of the consumer country concerned were satisfactory.

In order to mitigate the impact of such a right, INFCE suggested that the criteria for its exercise should be established, to the extent possible, before long-term fuel supply contracts are concluded or, for short-term contracts, before fuel is committed to nuclear reactors. It was also generally agreed that pending the development of common approaches to the exercise of the right of prior consent—as a first step toward broader international consensus—supplier countries should exercise that right in a manner that takes into account national policies and particular circumstances of consumer countries. The objective of this review should be to avoid, wherever possible, creating problems for a country's nuclear power program planning efforts. Subject to unusual circumstances, the right of prior consent should be exercised in a predictable manner that conforms to any understanding that may have been reached between the parties when the right of prior consent was established.

Institutional Arrangements

Both the seminar and INFCE agreed that institutional arrangements might have the best potential to improve the present nuclear nonproliferating system. They were seen as making important contributions to minimizing proliferation risks and to assuring supply. To this extent, supply arrangements, which have been already discussed above, are also included by INFCE in the package of institutional measures. The following additional measures were identified:

Back-up Arrangements. The first back-up arrangement was designated to be a uranium emergency safety network. It could build on existing ad hoc arrangements among utilities, primarily in Europe and the United States, and could be used for swapping or loaning fuel out of existing inventories for a limited period of time. It was considered that a more institutionalized form with some government involvement could be envisaged on the basis of this experience. Participating members (utilities and/or states) would commit an agreed portion of their existing stockpiles to a pool to be drawn on in case of failure of supply in accordance with agreed terms and procedures. Although this could probably begin on a limited scale, the experience gained in Western Europe and the United States suggests that there might be virtue in developing more systematic back-up arrangements, beginning at the utility level and gradually evolving into regional or even worldwide arrangements capable of meeting not only minor but major supply interruptions. In the case of such a development, there could be a network of pools, involving consumers and/or consumer countries, producers and/or producing countries, and various combinations thereof.

An alternative could be an international nuclear fuel bank. This bank would be made up of supplier and consumer states and would itself hold a stockpile of natural and low enriched uranium or claims to such uranium. These assets could be made available to a consumer state whose supplies were interrupted by a contract default that was not the result of a breach of its nonproliferation undertakings. However, it should be noted that such measures would be rather complementary or would be "by-products of the mistrust," as they were called by Dr. Lantzke at the seminar, and would serve only as a confidence medium for concerns about supply reliability.

Multinational Facilities. Multinational or international facilities could form an attractive institution. For example, they could provide easier safeguards implementation by making unauthorized operations more difficult. Moreover, such large facilities could reduce the number of smaller facilities.

However, both the seminar and INFCE took a cautious attitude toward the feasibility of multinational facilities. Such facilities would have a generally more complicated management system than national facilities. They could increase the risk of transfer of sensitive technologies and could create difficulties of coordinating physical protection and safety measures with those of the host country. Although, idealistically, it seems desirable that the evolution of institutional arrangements should be toward multinational ventures that could eventually result in the development of regional nuclear fuel cycle centers, the practical difficulties in establishing and operating such ventures should not be underestimated.

The idea of an international nuclear fuel authority (INFA) was among the various institutional models discussed. This idea was more or less rejected, since at present there is little indication of any demand for the services of such an authority and even less indication that a major international initiative could be undertaken to form and operate such an authority.

In connection with all international or multinational arrangements, it was recognized that decisions would be required on such sensitive questions as membership, financing, voting arrangements, conditions of access, dispute settlement, status of the host government, and the like. It was noted that a solution would have to be found for avoiding possible interference by the host government. It was also recognized that in addition to the multinational and international nonproliferation undertakings, some supplier governments require the conclusion of bilateral agreements or evidence that, when exported, the material or equipment will fall under another recognized bilateral agreement. The seminar concluded that existing examples in Europe, such as EURODIF and URENCO, both joint ventures in the field of enrichment, indicate that such arrangements could work when there is a strong commercial interest on the part of at least one partner, with the other partner joining in on a cooperative basis.

INFCE noted coincidentally to the seminar that the development of institutional arrangements must be viewed as a process of gradual evolution. Initially it seems likely that if reprocessing is carried out, the economic incentive to build large-scale plants may mean that it is not necessary for all countries to establish the technology simultaneously. Those countries that do build large national reprocessing plants could offer reprocessing services to countries that are at an earlier stage of nuclear development. As their nuclear programs and experience build up, these countries, in turn, could also be expected to progress to where reprocessing within their own countries would be justified by their nuclear programs. These early stage countries may then wish to consider the construction of their own plants and, in turn, be in a position to provide services to other countries.

As far as the back end of the fuel cycle is concerned, INFCE was of the opinion that the feasibility of an international mechanism for providing assurances to participating states concerning access to and management of their spent fuel, consistent with nonproliferation objectives, is worth investigating. Consideration should be given to international cooperation in developing spent fuel storage and management options. There is a need for such schemes, since there are countries that do not have, and do not plan to have, all the steps of the back end of the fuel cycle within their national borders. Thus, they depend on foreign nuclear industries and services. International spent fuel management schemes might improve prospects for storing spent fuel in general and thereby assist such countries in the economic and management aspects of spent fuel storage. Since the negotiation and implementation of multinational or international enterprises generally take time, it seems that for the near future national facilities will be the most realistic solution to avoid a deficiency in spent fuel storage. Moreover, the fundamental question remains: To what extent would individual countries be willing to offer sites and accept the agreed international conditions?

The specific question of the possibilities for international control of separated plutonium and arrangements for international storage of plutonium was also considered. The IAEA statute, Art. XII A (5), provides a possible basis for such an arrangement with respect to excess plutonium. It was concluded that a scheme for international storage of plutonium could have important nonproliferation and assurance of supply advantages. Whether or not most countries decide to reprocess their spent fuel, the scheme would be relevant, since separated plutonium already exists in the world and some countries have definite plans to continue to introduce reprocessing.

Export Control of Sensitive Equipment. INFCE took a very cautious attitude toward the export control of sensitive equipment and technology. This reflects the general negative position of most countries—particularly the developing countries, who would see a constant discrimination in such a practice.

The seminar was aware of the permanent discrimination of such a policy and was reminded of it by René Foch, who observed (see Chapter 6) that nonproliferation must be discriminatory almost by definition. Nondiscrimination would lead to an unacceptable lowest common denominator result. It was concluded that future sensitive facilities should be phased in only when economically justified and only in a multilateral framework. While this process may involve the transfer of sensitive technologies into areas of possible political instability, it was considered as the most effective method of controlling the inevitable.

FUTURE PROSPECTS

According to Robert Schaetzel (Chapter 7), nonproliferation is almost a laboratory case for international cooperation. Almost no other subject of international policy contains so many multiple and frequently conflicting goals—for example, nonproliferation versus energy supply, international versus national arrangements, environment versus energy, regional versus international arrangements, universal norms versus case-by-case approach, the goals of oil deficient consumer countries versus those of raw material producers, and the commercial inefficiency of international arrangements compared with those coming from a free market system.

However, although there are many difficulties, there are also many opportunities to overcome the problems. There is a general consensus opposing nuclear proliferation among both industrial and developing nonnuclear weapon states, for the risks are global and affect all countries. On the one hand there is the risk of atomic bombs and their use, and on the other there is the risk of energy shortages—both having potential diastrous consequences for humanity. At present the number of nuclear weapon states is few. There is a common interest between the United States and the USSR in the area. There is already a high degree of international cooperation (e.g., NPT), and opportunities exist to control critical material and facilities (e.g., the IAEA safeguard system). Finally, there is already a marked degree of international dependence within the nuclear industry, which could in itself be a catalyst to a future solution.

There is agreement about the goal to limit nuclear proliferation but not about the means to achieve it. The primary task therefore is to find the most realistic and promising means to reach this goal. The knowledge, technical capacity, and basic materials for making nuclear explosives are today quite widespread in the world. International cooperation is therefore indispensable to any policies aimed at minimizing the effect and preventing further proliferation. The development of a cooperative approach requires the participation of all states with a serious interest in nuclear energy development rather than confrontation between supplier and consumer. The lack of cooperation in nuclear power would make it even harder to achieve the much needed coordination in oil supply.

INFCE did not reconcile the different approaches but considerably improved the conditions for a future reconciliation. The United States and other supplier countries are more aware of the different energy situations of other countries that sometimes require other developments—for example, the breeder. The consumer countries are now more aware of the whole nonproliferation problem, which was often underestimated, particularly in Europe. INFCE did not reveal completely new facts or data, nor did it find surprise solutions to the various problems. However, in addition to improving the climate within the international nuclear community, which in 1977 was in disarray, INFCE produced a common factual background and findings that could and should facilitate a new international understanding at the political level. These routes must now be exploited. What technical experts evaluated during two years now needs to be implemented by diplomats and decisionmakers—hopefully in the near future.

What does this mean in practical terms? Countries should resist the temptation to seek unilateral solutions. This could be facilitated by the fact that no country, including the United States, has control over sensitive nuclear technology or over world uranium resources, either for purposes of supply or denial. At INFCE there has been little acceptance of the requirements by some supplier states that agreements for nuclear cooperation—for example, with the United States—be contingent on its having a veto over consumer fuel cycle activities. There was a belief that vetoes were likely to be counterproductive. By raising doubts about assurance of fuel supply, such policies will serve as stimuli to some countries to strive for independent fuel cycle capabilities, including their own enrichment and reprocessing plants.

It seems to be inconsistent to try to reduce the impetus for nations to acquire reprocessing and enrichment capabilities by increasing the assurance of uranium and enrichment supply, while conditioning such policies on consumer nations foregoing undesirable activities. While such policies could be effective in the short term, they could be counterproductive from a larger term point of view. There are signs[8] that INFCE has produced a reconsideration of such a policy. This will be revealed if the negotiations of present cooperation agreements take place. Such negotiations were suspended during INFCE, and the deadlines both for the United States–Europe (Euratom)–Japan and the Canada–Europe–Japan agreements have recently been extended to the spring of 1981. It would be interesting to see whether some features of the nonproliferation undertakings and criteria that were formulated at INFCE were found in these agreements.

Institutional arrangements seem to be the most promising means of strengthening the effectiveness of international cooperation in limiting proliferation. However, as both INFCE and the seminar pointed out, they will not come overnight, but have to be built up gradually. Among the most important elements is the strengthening of the IAEA safeguard system, including continuous international oversight of facilities with weapons-usable plutonium and highly enriched uranium. Safeguards must ensure that there is a fair chance of the would-be diverter being promptly caught. Attempts should be made to extend the present nonproliferation regime to those countries having significant unsafeguarded facilities—namely, Israel, India, and South Africa—as well as to condition all nuclear commerce on the acceptance of international safeguards on all civil nuclear facilities and weapons-usable material (full scope safeguards). The recent experience with the German–Canadian competition to supply Argentina with a heavy water reactor shows that opinions on this issue still differ (Germany, which got the contract, did not impose such a full scope safeguards clause, but similar conditions).

Moreover, internationally agreed upon sanctions, including those not related to nuclear matters, could be a threat to nations acquiring weapons through programs specifically designed for that purpose. This could be a means to counter clandestine proliferation, a problem that has often been neglected in recent debates.

As far as the multinationalization or internationalization of the fuel cycle is concerned, one has to face realities. Beginning with, or making maximum use of, existing national facilities and institutes and expanding from such a basis could be a starting point.

At present there are enrichment plants in operation or planned, in addition to those of the nuclear weapon states, in the Federal Republic of Germany, the Netherlands, Japan, India, Pakistan, and South Africa. There are already cooperative schemes in the FRG, the Netherlands, and the United Kingdom (URENCO) and in France (EURODIF)—the latter being joint ventures under the control of France. It is obvious that a sort of internationalization of plants in the remaining countries, while presently not achievable, is worth being explored—for example, in Japan.

There are reprocessing plants in operation or planned in the United Kingdom, France, Japan, the Federal Republic of Germany, and India. There might be a slight possibility for multinationalization of the reprocessing plants planned in Japan and Germany, at least offering access to other countries so that there is no attempt by countries with smaller nuclear programs to build their own plants. However, national legal requirements and local opposition sometimes reveal major constraints in such projects. Within a package of nonproliferation undertakings, the international nuclear community could further agree to limit the spread of weapons-usable plutonium and highly enriched uranium to economically justifiable projects including multinational facilities as dictated by economy of scale.

As far as the back end of the fuel cycle is concerned, the prospects for multinational solutions might be brighter. Centralized facilities for the disposal of spent fuel and/or vitrified high-level waste would alleviate the concerns of countries with small nuclear programs or those where suitable sites may not exist. These facilities—whether international, multinational, or national with access by foreign countries—would benefit from economies of size and could also reduce the diversion risk, since obviously a large number of spent fuel repositories would enhance the risk. Possibilities and feasibility of international spent fuel management are currently being discussed by an international expert group in the framework of IAEA.

The most advanced international project is the proposed international plutonium storage (IPS) regime, which is also being discussed by a group of experts under IAEA auspice. Such a regime would put national stocks of plutonium in excess of immediate needs under international control, probably under the control of IAEA. Although some practical and political difficulties have to be solved, the prospects of such a regime look good. This could therefore be a first step toward the possible creation of a broader regime providing for internationalization of other sensitive nuclear energy activities in certain circumstances. At the INFCE Final Plenary Conference all countries supported—at least verbally—such a regime. However, whether and how such solutions are applied will be a consequence of just how seriously the matter of nonproliferation is taken by the countries involved and the extent to which they perform as a political force in putting cooperative solutions into place.

The post-INFCE debate will take place primarily within the framework of the IAEA. The debate will be complemented by bilateral talks in the renegotiation process of nuclear agreements. In addition to the two expert groups mentioned above there are plans to establish a third group within the IAEA that will be formed to investigate assurance of supply questions. The speed and the quality of these discussions will determine whether this will be sufficient to keep the good cooperative climate of INFCE alive and whether the discussion can transform the spirit of INFCE into practical solutions that would improve the present global nonproliferation scheme.

Both the Salzburg Seminar and INFCE stated that there will be no new universal approaches or solutions. However, there are many elements that contribute to gradual solutions. All solutions require international cooperation. In this sense, "internationalization" is the only alternative to nuclear nonproliferation in our lifetime.

NOTES

1. Non-Proliferation Treaty, art. IV.2.
2. The unsafeguarded plants in India consist of two fuel fabrication plants, three reactors, and two reprocessing plants. In Israel there is a large research reactor and a pilot reprocessing facility; in South Africa, a pilot enrichment plant; in Egypt, a small unsafeguarded research reactor; and in Spain, an unsafeguarded power reactor operated jointly with France.
3. INFCE, Summary Volume, p. 276, published by the IAEA (Vienna 1980).
4. See also Pierre Lellouche, "Non Proliferation after INFCE: Risks and Opportunities," Foreign Affairs, October 1979. (Mr. Lellouche participated in the Salzburg Seminar.)
5. *INFCE Summary Volume* (Vienna: IAEA, 1980), p. 37."
6. Ibid., p. 36.
7. Ibid.
8. For example, Gerard Smith, U.S. ambassador at large for nonproliferation matters, said in his statement at the Final Plenary Conference of INFCE in February 1980: "We are now more conscious of the long-range implications of near-term actions" (Wireless from Washington (USICA Bonn) No. 39, 26.2. 1980). See also Gerard Smith and George U. Rathgens, "Nuclear Energy and Non-Proliferation in the Wake of INFCE" *Trialogue* 22 (Winter 1980).

Seminar Reports

POLITICAL ASPECTS: REPORT BY PROFESSORS SCHAETZEL AND FOCH'S SEMINAR GROUP

Basic Premises

Nonproliferation policies have to be viewed in the classic dilemma between "idealism" and "Realpolitik." One of the basic premises was that no radical new approach will solve the proliferation problem given the current political economic and technological context both at a global scale and also from region to region and country to country. Furthermore, possible policies themselves are limited: this is particularly the case of "technical fixes," which alone cannot provide a solution to proliferation.

The second premise was that the current nonproliferation regime, with its various components (i.e., IAEA, NPT, London Suppliers Group, INFCE, and other bilateral arrangements), is probably the best that can be achieved given present conditions. This regime is nevertheless only modestly successful, given its inherent discrimination components.

Improvements: Aim and Conditions

Looking at possible ways to improve the current nonproliferation regime, the seminar emphasized that efforts should be aimed at achieving new codes of conduct and institutional arrangements that will be as simple as possible and relatively narrow, so as to be both effective and attractive to prospective participants. In general, universal approaches were found to be unsatisfactory because they seemed to lead to meaningless or inappropriate solutions to important cases, as well as being almost impossible to achieve.

Multinational Fuel Cycle Management as a Possible Improvement of the Nonproliferation Regime

There was agreement that security concerns are a critical factor in a government's consideration of the nuclear option. The major components of this security question were identified as (1) military and (2) energy-economic factors. Multinational arrangements only answer the economic-energy dimension of the security problem, while leaving unresolved the fundamental question of nations' military security.

On a general level, the seminar agreed that the international community might help countries address their economic-energy and security concerns through a multilateral approach to certain parts of the fuel cycle. Multilateralized facilities could help meet countries' legitimate energy needs, while at the same time—on balance—helping to discourage proliferation.

It was concluded, however, that there are definite limits as to what parts of the fuel cycle should be multilateralized and as to under what conditions this process would succeed. A "geographical grouping" approach was judged to be the most effective method of addressing security concerns, with regional centers established only in areas where there was a clear economic justification based on a need for services. The existing examples of EURODIF, and URENCO indicate that such arrangements work when there is a strong commercial interest evident on the part of at least one partner with the others joining in on a cooperative basis.

The seminar also discussed the problem of siting, in which commercial factors, national interests, and international considerations all play a role. Most members of the seminar felt that Europe was the best starting point for the establishment of such arrangements, because of its present and future nuclear power base.

More specifically, the seminar was in general agreement that further development of reprocessing and enrichment and corresponding institutional arrangements must await the outcome of INFCE. However, further feasibility studies for these centers by interested countries should proceed prior to INFCE's formal conclusion in 1980. Plutonium storage under a multilateral framework was supported as probably the best method of avoiding seizure or diversion of large amounts of plutonium, provided that the storage be located in a secure place (e.g., a nuclear weapons state).

The seminar considered the needs of developing countries within a multilateral framework. The seminar was divided on the immediacy of the need of nuclear power in the majority of developing countries. In view of the fact that denial of such sensitive technology is impossible in the long term, especially to the most advanced of the developing countries, it was agreed that the establishment of multilateral arrangements should be supported by supplier nations. The group decided that sensitive facilities in developing regions (and elsewhere) should be phased in only when economically justified and only in a multilateral framework. While this process may involve the transfer of sensitive technologies into areas of possible political instability, it was considered to be the most effective method of controlling the inevitable.

In conclusion, the seminar's answer to the basic interrogation (i.e., internationalization as an alternative to proliferation) is a modest "yes." In this respect, the steps toward multilateralization of certain parts of the fuel cycle that the seminar agreed on are rather modest efforts that should improve nations' feelings of economic-energy security and complement technological efforts to deter proliferation.

Other Questions

Other topics were covered by the seminar in a more random manner. These included the question of incentives and disincentives to the states' participation in new nonproliferation efforts, as well as the siting and administrative control of an international fuel bank. The seminar also discussed the historical evolution of EURATOM–U.S. cooperation, with some seminar members believing that the past course of this relationship persists in today's situation and others asserting that circumstances were considerably altered. Finally, the seminar considered in greater detail the question of "vertical" versus "horizontal" proliferation as well as the problem of sanctions against potential proliferators.

With respect to the first item, some members of the seminar strongly believed that there is a direct link between nuclear weapons states' disarming according to the spirit of article VI of the NPT and nonnuclear weapons states' willingness to forego such weapons. Other members thought that there was very little linkage, because the individual country's security perceptions and needs were more significant determinants of the nuclear option than global arms control.

Regarding sanctions against potential proliferators, the seminar decided that while a universal code of conduct was intellectually attractive, the norms of "democratic practices" currently existing in international bodies would make it impossible to agree whether an infraction had occurred or on what penalty should be imposed. In the context of limited multilateral arrangements, as previously mentioned, multilateral sanctions could, however, be useful when applied by the various participants. Where the offending country is not a party to such multilateral arrangements, bilateral sanctions could, on the other hand, be the only effective form of discussion. On a more general plane, some of the seminar members thought that it might be possible to work out criteria to govern application of bilateral sanctions, but others felt that an explicit list of such measures would be undesirably rigid and probably difficult to work out.

TECHNICAL ASPECTS: REPORT BY PROFESSOR AGNEW'S SEMINAR GROUP

Basic Premises

The technical background information provided by Professors Seaborg and Agnew on which the discussions were based indicated that there are four possibly proliferation-resistant fuel cycles:

1. The once through cycle, which means "do not reprocess."
2. The denatured U-233–Thorium cycle.
3. The CIVEX cycle. The idea here is to keep enough fission products in the recovered fissile material so that its usage becomes difficult and dangerous.
4. The denatured Pu-239–Uranium cycle.

The last three alternative cycles are apparently feasible, but two major questions remain still unanswered: How much will it cost? How long will it take to develop one of these methods on an industrial scale?

Technical Fixes

It was clear from the observations made during the group sessions that there are no technical fixes to prevent nuclear proliferation. The only thing that can be done is to make it technically more difficult for nations to divert nuclear materials from peaceful uses toward nuclear weapons. The major obstacle to making a bomb is the lack of fissile material; the main problem is that of safeguards and physical protection and not of technical options. In this respect, a comparison between various enrichment processes was made. From the viewpoint of proliferation, the gaseous diffusion process presents very little danger. As to the centrifuge and laser processes, a lot of work is still to be done, but a future proliferation problem could arise. As far as seizure by terrorists is concerned, it is obvious that one should inhibit access to fissile material. However, it should be pointed out that it is not simple for a small group of individuals to build a bomb. Even once fissile material has been obtained, a high degree of technological sophistication and access to the appropriate technical facilities are still needed. The seminar concluded that possible technical improvements of the fuel cycle will not suffice to prevent proliferation. They could only stop terrorists, but would not deter a government which is absolutely determined to construct a nuclear explosive device—especially since the reliability of such a device does not need to be proven by testing.

Political Fixes

The proliferation problem thus calls essentially for political and psychological answers. Three more general questions arise in this context:

1. What could be the incentives for a state to go military-nuclear?
2. What are the political, technical, and economic means to inhibit these incentives?
3. How does the development of the different aspects of civilian nuclear programs affect the situation?

The answer to the first question could be summarized in the following six points:

1. Survival could be the first incentive. Such countries as Israel, South Africa, South Korea, Taiwan, and Pakistan were mentioned.
2. Prestige.
3. Regional leadership, the quest for military strength, "lust for power."
4. Security reasons.
5. An apparent move toward nuclear weapons could be made for the sake of gaining additional bargaining power in exchange for stopping that move.
6. Peaceful nuclear explosions could be an incentive or a pretext.

The following substitutes to each of these six incentives could be distinguished:

1. As far as prestige is concerned, one could provide prestige incentives in other areas. It must also be stressed that "nuclear prestige" can turn out to be a costly mirage (e.g., India).
2. For survival and security, action could be taken for a positive or negative "umbrella policy" by which the powers concerned could give or refrain from giving security guarantees to other countries. The case of South Korea was cited as an example. Interregional cooperation could, in some cases, serve the same ends.
3. No clear answer was proposed on the issue of dissuading a country engaged in a deliberate policy of attaining the status of most powerful nation in a given area.
4. The eventual benefits from the development of nuclear explosive devices should be shared. As pointed out by some members of the group, they should be available to all NPT partners and to all countries accepting IAEA safeguards.
5. Finally, some members of the group proposed that effective measures should be taken in the direction of nuclear disarmament as a real disincentive against nuclear proliferation.

The group came to the following conclusion concerning the affect of civilian nuclear programmes on the situation: Past experiences confirm that it is not necessary to have an industrial nuclear power program to acquire nuclear explosive devices. However, the presence of a nuclear power program could enhance a nation's potential ability to develop a nuclear explosive capability. In this respect, the group considers that attention should therefore be focused more particularly on the disposition of spent fuel.

INDUSTRIAL AND SOCIAL ASPECTS: REPORT BY PROFESSOR PORTER'S SEMINAR GROUP

Basic Premises

There seems to be some ambiguity with the theme of the seminar—Is internationalization the alternative to nuclear proliferation? That does not seem to be an alternative. International control exists and must remain. Incidentally, one question that has been inadvertently discussed at length is, Is Americanization the alternative to nuclear proliferation?

And if internationalization is a difficult concept, proliferation is equally ambiguous. Proliferation used to be shorthand among the cognoscenti for the spread of nuclear weapons, but now there is a strong tendency for the term to embrace all sensitive nuclear technologies. The Indian explosion has raised the question of whether a "peaceful explosion" is in fact "proliferation with intent—intent to frighten one's enemies." There no longer seems to be a simple litmus test for proliferation, and it looks as if the world is going to develop a case-by-case assessment of nuclear developments based on the capacity and the intentions of the country concerned.

The Relevance of the Nuclear Industry to the Proliferation of Bombs

Despite the almost total concentration in the seminar on the power industry, it was felt that it is only a minor part of a nonproliferation policy that should use all conventional means of power, diplomacy, and institutions. In any case, it seems that a nuclear power reactor is not the ideal route for making a bomb; instead it is a slow and uneconomic method. However, the seminar was aware of the dangers and agreed that any nuclear reactor—be it a power plant or a research facility—carries proliferation risks and any new member of the nuclear family is a potential threat. Under the umbrella of nuclear development know-how is acquired that could be useful in weapon making, and the technical risks cannot be confined to sensitive technologies of enrichment and reprocessing. These simply contain the highest risks for proliferation.

Once in existence, even if built with the best of good intentions, nuclear facilities constitute a permanent temptation for weapons use, especially if a country's political situation changes for the worse. Moreover, their very existence would always give a country a bargaining counter—essentially the simple threat that "unless someone solves our political or military problem, we'll use our plutonium to build a bomb."

Current American Policy, Aims, and Results

During the seminar, European nationalism has sometimes come to the fore, so our group would like to start by congratulating the Americans for generating wider interest and greater sensitivity to these problems. They have given a new impetus to the long-term strategy for nuclear development as well as giving proper prominence to the special dangers of sensitive technologies. However, the seminar was critical of American tactics and of the fact that too much of their policy is based upon the American energy situation and their near monopoly of the enriched uranium market. Their Non-Proliferation Act had some of the features of a preemptive strike: they legislated first, before they had general acceptance from their nuclear partners on the necessity to change.

Overall American policy, including the cancellation of their own breeder, has generated a fear that the nuclear future will be circumscribed and the options reduced. On the breeder question alone, Europe and Japan appear to be out of step with American technology and that may have serious technical and commercial disadvantages. One other result is that the plausibility of even the most nationalistic advocates of nuclear independence is increased. By and large, the seminar thought that the idea of national nuclear independence is a myth, but even so, it may still become the goal of some—to the disadvantage of all. In the short term, American policy has undermined confidence in America as a reliable supplier of fuel.

There is the more abstruse point that by extending the meaning of proliferation, America has shaken confidence in existing nonproliferation accords. Those accords were generally accepted: now that they are in jeopardy, it could lead to greater ambiguity and distrust. In the long run the whole debate—at times it has even been a confrontation—could lead to the strengthening of international nuclear institutions. So far, however, the positive side has been submerged in the intense controversy surrounding the rest.

INFCE and the Breeder

INFCE has provided a breathing space, time for the Euro-American controversy to die down. It may have some fruitful technical results in suggesting proliferation-resistant systems, but the seminar did not expect significant technical propositions to emerge in the two or three years spoken of. The scope of INFCE is much greater than that of any previous nuclear forum, and it might be a starting point for the permanent international coordination and control of nuclear technology and expertise. It could fill the gap left by IAEA, which has failed to keep the initiative in developing nuclear nonproliferation policy.

As to the breeder, American policy has cast a cloud over its development because it has questioned the need to build such a system and emphasized the dangers. They have little domestic need for the breeder, in contrast to the official French, Japanese, and German assessments of their own national positions. All three are very short of indigenous supplies of conventional energy, and they all resent their energy dependence. And the desire to be independent always has much political force. Nevertheless, the economics of the breeder will depend to a small degree on the price of uranium and to a much larger degree on the success and cost of the actual system developed. But however, due to the cost involved, the seminar is convinced that breeders will be very unattractive to any but the most advanced nuclear countries.

The Public Nuclear Controversy

There is the likelihood that the nuclear industry will become the symbol of all that is nasty in industrial society. Some of the seminar regard that as unfortunate, while others regard that as only a statement of the truth. It is interesting that in Europe so far, the chief issues to concern the public have been safety and waste management. The proliferation of bombs has faded somewhat from public consciousness, and the proliferation of plants that produce weapons material has barely entered the field of public debate. But public consciousness will grow, and public resistance will grow far beyond the skirmishes that have frightened the industry so far.

It was said that public opinion is a real problem, but it is rather the nuclear industry that is the problem. The industry is there to serve and not the other way round. If the industry and its institutions with all their resources cannot get public acceptance, then either they are incompetent or they do not deserve to.

The seminar's final philosophical point is that nuclear energy is not a virtue in itself. It is simply a necessity that has always to be looked at within the framework of other broader policies—energy, the environment, health, safety, and security.

Summary of the Seminar Discussion

*Ulf Lantzke**

SIX FACTS OF LIFE

In considering the problem of nuclear nonproliferation we have to face six facts of life as regards nuclear energy policy:

1. Nuclear power's contribution to energy supply is indispensable given the overall world energy supply and demand situation and its future prospects. If a general nuclear moratorium should occur, grave economic and social consequences would be unavoidable. The overwhelming majority of governments in the East and in the West, in the North and the South, are agreed on this point, and one must accept that these governments are fully aware of their responsibilities. Even if a given country were to decide against the use of nuclear, overall nuclear development should not be held up, since the economic advantages to other countries are large in relation to their needs. Delaying nuclear even temporarily is also not advisable, since a rather hasty development later on would endanger orderly, controlled development.
2. Nuclear weapon states and potential future nuclear weapon states are a reality.

*Executive Director, International Energy Agency, Paris

3. Nuclear technology is known worldwide and is already fairly widespread. Further diversification of nuclear knowledge is to be expected. There is no way of preventing people from improving their technological potential. History is obvious proof of that. The policy of denial has not stopped other countries from achieving technology. On the contrary, in some cases, it has motivated countries to develop weapon capability. At present, about fifteen countries have the capability of reprocessing and about ten countries of enriching uranium, and the number will grow. We will have to live with this situation, and we will have to tailor possible solutions for the real world, not a perceived one.
4. It must be recognized that the most economic and technically most feasible way to produce a nuclear weapon is not through the by-product of peaceful applications but rather through specialized technical equipment—for example, "research reactors" or, rather, "plutonium factories." Nevertheless, it must also be admitted that the peaceful use of nuclear energy produces as by-products material that can be turned, although with some difficulty, into weapons material. Massive peaceful use of nuclear energy can thus multiply opportunities. Quantity could turn into quality.
5. The diffused public perception of nuclear energy problems has developed into a real constraint. The public has not been well informed and is in many cases misinformed. Public education is a difficult task, and we cannot bring everybody for two weeks to Schloss Leopoldskron. We cannot even force the public to listen and believe; but the public perception of nuclear energy remains a very clear challenge to political leaders. They must exercise leadership, encourage better public education, and resist the temptation to follow misinformed public trends. The confidence of the public in institutions is an important issue and is a precondition for greater public acceptance of nuclear.
6. There is no way to exclude LDCs from the benefits of nuclear energy, especially if a nuclear program is economically viable and justified. Nuclear technology for the sake of prestige should be avoided. However, this is a difficult and delicate issue. On the other hand, we have to take into account that oil-exporting countries must be able to look with confidence to the postoil era and that developing countries in the stage of economic takeoff must have the energy means to broaden their economic base.

THE COMPLEXITY OF PROBLEMS

We have learned that proliferation problems are very complex. A variety of attitudes have been expressed throughout the seminar by some of the most prominent and knowledgeable experts on the subject. There were practically no polemics, and the differences were not as large as expected. I think that there was more or less a common sense that the proliferation problem is a very real one requiring rethinking and reorientation on a worldwide basis. But we must bear in mind that we are not dealing with only one problem but rather with an extremely complex set of issues.

First, a suitable definition of proliferation is necessary. (1) Is it proliferation of any nuclear activity? For example, would a widening of nuclear knowledge be considered a proliferation threat? (2) Is it the enlargement of physical availability of weapons-grade material, concentrating on U-233, U-235, and Pu-239 (in pure form or in diluted more difficult form)? (3) Is it the proliferation of knowhow, of ability to become a weapon state?

Most attention is directed to question 2. Although it is generally accepted as a sound premise and one where improvement would be useful, it really does not cover all of our subject. The proliferation of nuclear activity (question 1) is one of the facts of life and, therefore, not possible to prevent. Delays of the proliferation of know-how (question 3) might be possible but probably not advisable. Therefore, more efforts should be made to improve the overall climate to be more decisive. I also wonder whether it is really possible to control physical material to such a degree as to prevent bomb building if technological know-how is available to a state.

Second, what are the real sources of danger? Two groups have been identified—states and terrorists. This is a strange grouping—certainly two totally different types of problems in quality. Terrorists problems seem to be manageable by adequate physical protection. (In addition, the difficulties in developing an effective weapon would also serve as a constraint for this group.) Thus we are left with the problem of finding means that make it more difficult for states to divert or seize weapons-useable material.

There is a whole range of psychological problems of a different nature: the most acute are concerns about availability of the nuclear option for the economic and social well-being of a given economy. This highly sensitive issue has created a period of mistrust with clearly counterproductive effects. Concerns of countries that they may be maneuvered into a second rate economic position as well as into a politically inferior position should be taken seriously. Therefore, there is a need to adhere to stable and reliable conditions for nuclear trade. Even if one does not accept these concerns as being justified, they are real and have a direct impact on reactions concerning the main problems, and it shows that presentational diplomatic handling of issues is as important as substance.

The motivations to develop nuclear weapons are mainly threefold: (1) prestige, (2) desire to gain weight in international position (although doubtful), and (3) real concerns about security and conception or misconception of improving security by having a nuclear weapon. And finally, there is an interaction of government activities and public reaction. If governments overstress their concerns, there is a counterproductive influence on public understanding. One should not forget the role media may play in spreading basic fears in a transparent worldwide society. Nor should one overlook the links between military security and economic security, which put the whole issue in a much broader context.

Finally, one has to consider the roles and relationship of industry and government. On the one hand, industry is clearly in favor of industrial management and responsibility and a competitive system for nuclear supplies. On the other hand, there is a clear government responsibility, given the sensitive nature of nuclear developments. Can we really accept the notion that the nuclear industry is just another industry? On the other hand, can we ignore the advantages gained by industrial management capabilities? Do we need to rethink the basic structural relations between industry and governments and to develop rules for this?

POSSIBLE SOLUTIONS

The discussion has shown that expectations of possible solutions should not be set too high. None of the speakers indicated that there is any absolute solution. The facts of life and the catalogue of problems illustrate why it is so difficult to make progress. On the other hand, this discussion has shown that there is sufficient time to search for solutions in an unemotional climate. There is a need to tackle problems at a given time without losing perspective for longer term developments.

Technical Solutions

There seem to be devices that would make it more difficult to use "waste" from reactors for weapons but they have their limitations with regard both to their high costs and to their effectiveness. Therefore, the question arises whether higher costs and less technical perfection will be accepted, given that proliferation results are limited. In any case, it is fair to assume that technology alone can give only very limited answers.

Existing Nonproliferation Systems

These include IAEA, Euratom, NPT, guidelines of London Nuclear Suppliers Club, and nuclear free zones such as in Latin America. What effect can they have? How can they be improved? Are full scope safeguards sufficient? These questions remain to be answered. However, it is clear that the international understanding developing from these organizations has built certain psychological barriers against going for weapons. Therefore, they represent precious capital that should not be wasted.

International Cooperative Management of Sensitive Parts of the Fuel Cycle

Such schemes are basically reasonable and achievable if political and economic objectives can be integrated. Some are already in existence but mainly for economic reasons. EURODIF and URENCO could serve as starting points in the enrichment sector since the by-product of cooperative management is broader control. Therefore, initiative should be taken to bring both security and economic objectives together.

As far as reprocessing is concerned, many small plants do not make economic sense. With respect to safety issues, one could imagine that plutonium storage and control, which has been successful in the military sector, could also be effective in the civilian sector if similar systems of physical protection are applied. Although the concept of an international fuel bank is a by-product of "mistrust" rather than of proliferation protection, it could serve as a confidence medium for concerns about supply reliability.

Other Areas

Stricter bilateral controls by suppliers, including leasing techniques, are highly sensitive—at least psychologically speaking. Counterproductive elements and possibilities for consumers to look for other opportunities have to be assessed. A need for a universal system becomes more and more obvious.

Does a go slow policy on the breeder and reprocessing development really help? Would not controlled development be more useful?

Some understanding about the "rules of the game" for industry are needed. Although competition is principally good, excessive competition involving speculation on the eagerness to be a nuclear state may be damaging and lead to unacceptable results. Industrial fears of competitive disadvantages are very strong, as the experience with getting the NPT accepted in Germany has shown.

Finally, with regard to INFCE, should we really question its usefulness? Are we doomed not to succeed in INFCE because a negative result would give possibly limited advantages to one country or the other but on the whole would be damaging to everybody, because the public reaction to such a negative outcome could be very counterproductive?

CONCLUSIONS

In closing, I would say that we have

1. To live up to the facts of life in nuclear developments;
2. To deemotionalize the debate and get down to hard facts;
3. To create a climate whereby developing weapons is more damaging to a nation's standing in the world than not developing weapons;
4. And finally, to bring together economic, political, safety, and proliferation aspects in an optimal way.

The answer to the question, Is internationalization the alternative to nonproliferation? would be "yes"—but a yes well understood.

Part II

Chapter 1

Nuclear as a Component of International Energy Supply

*Ulf Lantzke**

THE NEED FOR NUCLEAR

I think that it is important in this introductory chapter to illustrate why the nuclear issue is a vital economic issue for our societies. Simply put, the world today cannot do without nuclear power without causing severe harm to the basic foundations of our society. The indispensable role of nuclear power is readily apparent from an analysis of the medium- and long-term energy prospects of the Western industrial democracies and of other parts of the world. Since the 1973–1974 energy crisis, a large number of studies dealing with long-term energy trends—among them studies by the International Energy Agency based on the policies of its member countries and a separate, independent *World Energy Outlook* by the Combined Energy Staff of the IEA and OECD—have been published. Although they differ to varying extents as to their underlying assumptions, the basic conclusion of nearly all of these studies is that unless we strengthen our existing policies, a severe imbalance between potential energy demand and available energy supplies could develop as soon as the mid-1980s. This imbalance is best illustrated by an examination of world demand

*Executive Director, International Energy Agency, Paris

for OPEC oil, since the world oil market is still regarded by virtually all countries as the residual source of energy necessary to balance domestic demand and production.

Taking the 1978 IEA study of the energy policies and programs of our member countries, Table 1–1 shows, based on energy programs now in place, how the global oil balance will look by 1985.

It is this potential shortfall that signals loudly and clearly that we face a potential energy crisis in the 1980s. Of course, these projections are based upon a number of assumptions that may or may not turn out to be precisely accurate—assumptions about economic growth, prices, effectiveness of conservation, and supply programs. But even assuming a conjuncture of the most optimistic assumptions, the looming energy problem will at most be postponed for a few years unless we really implement a set of energy policy measures that are much stronger than governments and legislatures have been willing to contemplate so far. And remember too that these few years respite may be dearly bought in terms of lower economic growth or sudden and disruptive increases in energy prices that could severely strain the international economic system.

I do not want to suggest that an energy crisis, with its associated economic, and even political, damage, is inevitable. It can be avoided—if, and only if, we are prepared to give strong support to a wide range of forceful energy policy measures. Nor do I want to suggest that nothing has been done. Some progress has and is being made, but it is not yet enough to cope with the potential seriousness of the situation that we will face in the latter part of this century as oil becomes less and less available—first, because of production ceilings that may be imposed by some of the key producers; and second, because of physical limits as world oil production peaks and gradually declines.

We will have to take action on both sides of the energy demand and

Table 1–1. Projection of OPEC Oil Demand and Supply

	mbd
Oil import demand of IEA countries	25–30
Total non-IEA demand (includes non-IEA OECD and developing country oil imports, as well as OPEC internal consumption)	8–10
Projected demand for OPEC oil	33–40
Estimated OPEC production, taking into account both economic and political factors	30–35
Potential shortfall	3–5

Source: IEA, 1978.

supply equation in order to achieve a satisfactory balance. We simply cannot afford to neglect the contribution of any of the solutions proposed if we are to achieve essential social, economic, and political goals.

This means that we must pursue strong conservation policies to increase efficiency of energy use. We must embark on an aggressive coal utilization strategy and develop greater international coal trade. We will have to introduce new energy technologies as rapidly as possible on a commercial scale. And we cannot afford to neglect the vital contribution that nuclear power can make—particularly the realistic potential for fast growing nuclear energy supplies during the remaining years of this century. It must be recognized too that we face not only an oil problem but a total energy problem, because each of these solutions also contains limitations, and we have to rely upon a mixture of all of them.

Conservation potential and constraints vary widely among energy-using sectors and among countries. Many good housekeeping measures to eliminate obvious waste have been undertaken successfully since 1973. Making deep and lasting reductions in the amount of energy used per unit of product or activity is a longer term task. It will require the combined effects of energy prices, government action, changing social patterns and norms, structural changes within the economy, innumerable decisions by many users, large and small. Lead times can be very long—for example, up to a decade to change over a stock of cars, twenty to thirty years for most industrial equipment, a century or more for a nation's entire stock of housing. A continuing and intensive commitment is required over many areas that individually may appear to render only insignificant savings but taken together represent a substantial reduction in demand.

Fuel switching also offers potential, particularly in the key sectors of electricity generation, process heat in industry, and, in some countries, district heating. Nevertheless, limitations exist in the short term. Conversion of industrial processes and installation of district heating carry considerable financial costs. Most power generation capacity due to be commissioned for the 1980s has already been ordered, and some of this is new oil-fired capacity. Power station managers are reluctant to construct new coal-fired capacity because of reduced forecasts of electricity demand, uncertainty over the availability on a reliable basis of coal supplies, and environmental constraints. There also seems to be a growing reluctance on the part of utilities to buy a nuclear plant, if they have any choice, because of the procedural and public opposition headaches involved.

Coal is a realistic alternative because it is available in generous quantities and greater amounts of coal can be used, particularly for electricity generation, industrial process heat, and district heating systems. In the IEA we are trying to develop programs that will greatly expand coal use and trade. But there are limits to this alternative. Major and early investments are required for mines and transportation facilities. Many environmental, economic, and social issues need to be solved before significant mine expansion, increased coal use, and a greatly expanded coal trade would be acceptable. New coal-fired equipment and improved techniques for controlling air pollution will have to be provided without making coal noncompetitive with other fuels. Reliable long-term arrangements need to be developed between coal-importing and coal-exporting nations. The potential for coal in the near term appears to be constrained by these economic, technical, and environmental problems. In the longer term, with prompt government action to overcome these constraints, coal can make a major additional contribution to energy supply toward the late 1980s and the early 1990s.

Nor can we rely to a very great extent on the new energy technologies. Vigorous research and development efforts should be pursued into new energy sources such as solar energy, biomass conversion, and ocean and wind energy as well as nuclear fusion. In the IEA we have cooperative programs in all of these areas. However, they do not appear to offer any significant supplies much before the end of the century, let alone by 1985 and 1990. By all means we must continue to develop the possibilities that do exist, especially for the longer term, but it is not realistic to abandon nuclear power for the medium term on the strength of wildly optimistic expectations of new energy sources being available by then.

To concentrate on the role of nuclear power in meeting overall energy requirements, over the past twenty years rapid progress has been made in nuclear technology. Proven reactor systems have been in safe operation in the principal industrialized countries, although nuclear power has not as yet made a large contribution to overall energy supplies. Yet the share of nuclear is growing very rapidly, although admittedly from a very low base. In 1974 installed nuclear capacity in the Free World represented only some 85 million tons of oil equivalent (mtoe). In the IEA countries the growth in nuclear power has been quite impressive, growing from 41 mtoe in 1973 to 85 mtoe in 1976 and 120 mtoe in 1978.

But despite these indications of growing use of nuclear power, serious problems have surfaced, causing significant delays in nuclear power plans. Unless there is a determined international effort to address both near- and long-term issues concerning nuclear energy within their respective time frames, further serious shortfalls in nuclear power projections could occur—shortfalls that the industrial nations as a whole simply cannot afford. For even the most strenuous efforts at conservation and expansion of indigenous oil, natural gas, and coal supplies will not be sufficient to fill the energy gap of the 1980s and beyond without a substantial increase in nuclear power.

Recently, estimates of member countries' installed nuclear capacity by 1985 have again been revised sharply downwards, due to uncertainty over future growth in electricity demand and the rapidly intensifying public debate on nuclear power, both of which restrain or delay the construction of nuclear power stations.

Despite the outstanding safety record of operation of nuclear power reactors over the past twenty years, increasing concern has developed, in particular as regards the dangers posed by the transport, storage, and treatment of spent fuel elements and the disposal of the resulting highly radioactive wastes. The reprocessing of spent fuel to recycle plutonium and uranium and the development of the fast breeder reactor also raise important concerns about the risk of nuclear proliferation. Decisions about nuclear power require that effective solutions to these problems be reached on an urgent basis. In the near term, decisions will have to be made on the construction of new nuclear reactors, expanded uranium exploration and mining, stable international arrangements for supplying nuclear fuels and technologies, and development of facilities for uranium enrichment. Further intensive research and development on waste disposal will have to be carried out within the lead time of construction of new nuclear reactors.

THE IEA AND NUCLEAR

The governments of the industrial nations do have a forum for coordinating their energy policies—the International Energy Agency. The IEA has a membership of nineteen countries and was formed after the energy crisis of 1973–1974. From its beginnings in 1974, it has taken a positive attitude toward nuclear power. This is explicitly recognized in Chapter VII of the Agreement on an International Energy Programme, which led to the creation of the IEA. Particular attention is paid to the need to develop nuclear energy and to ensure adequate supply of natural uranium resources and enrichment services. The first energy R&D projects of the IEA also included cooperative action on nuclear technology in the field of nuclear safety, designed to encourage active nuclear development as part of the overall effort to accelerate development of alternative energies to oil.

With the growing realization that an all-out effort must be made to solve the energy problem, the IEA has adopted a strongly positive attitude in favor of an expanded nuclear contribution. At the meeting of IEA ministers that took place October, 1977 an objective was agreed to hold net oil imports to not more than 26 mbd in 1985 compared to about 22 mbd at present. Basic principles for energy policy were endorsed that included the following policy approach to nuclear power:

> Steady expansion of nuclear generating capacity as a main and indispensable element in attaining the group objectives, consistent with safety, environmental and security standards satisfactory to the countries concerned and with the need to prevent the proliferation of nuclear weapons. In order to provide for this expansion it will be necessary through co-operation to assure reliable availability of:
> —adequate supplies of nuclear fuel (uranium and enrichment capacity) at equitable prices;
> —adequate facilities and techniques for development of nuclear electricity generation, for dealing with spent fuel, for waste management, and for overall handling of the back end of the nuclear fuel cycle.[1]

The problem of radioactive spent fuel and waste disposal for the entire nuclear fuel cycle is rapidly emerging in some countries as a main impediment to nuclear development. Until there is more progress toward a satisfactory solution, some governments will hesitate to approve new nuclear power projects or may even be unwilling to do so. It appears that some governments have overestimated the preparedness of countries undertaking reprocessing to store nuclear waste from other countries. Some countries with reprocessing facilities are moving toward contractual arrangements under which countries sending fuel for reprocessing might be asked to accept the return of long-lived waste arising from such reprocessing. In the case of the United States, while acceptable long-term solutions are being sought during the reprocessing moratorium, and the United States has indicated a willingness in this context to accept foreign spent fuel on a limited basis, it is by no means clear that the United States would be prepared to make unconditional and open-ended commitments for waste management and disposal. It is clear, therefore, that efforts at devising a realistic solution to the full range of issues must be intensified.

The IEA is currently active in promoting government action to find solutions to nuclear problems. The IEA nuclear work program covers all stages of the nuclear fuel cycle. It is being implemented in close liaison with other complementary programs being undertaken in other international bodies such as the IAEA, the OECD's Nuclear Energy

Agency, and the International Nuclear Fuel Cycle Evaluation (INFCE). Emphasis is laid on a continuing assessment of future electricity growth and the required share of nuclear power as well as the conditions governing supply of natural and enriched uranium and arrangements for the back end of the nuclear fuel cycle. Some of these activities are closely akin to examinations underway at INFCE in an effort to develop possible technical solutions, including a clearer international perception of options to resolve problems connected with the entire nuclear fuel cycle. Accordingly, the IEA is participating in some of the INFCE work and monitoring its progress.

THE WAY AHEAD

Although optimistic projects for further rapid growth of nuclear have recently been scaled downwards, current anticipated growth is still quite striking. For example, the 1978 IEA review of member countries' current energy policies estimated nuclear power production at 169 mtoe in 1980, 352 mtoe in 1985, and 587 mtoe in 1990—a sevenfold increase in the fourteen years from 1976 to 1990 or an annual average growth rate of almost 15 percent. Although it must be recognized that much of this additional nuclear energy production will be at risk unless government policies are implemented with strong resolve, these figures illustrate that most industrial countries count on nuclear power to meet a growing share of their energy requirements.

Nuclear power production in IEA countries is anticipated to meet over 11 percent of total energy demand in 1990 compared to less than 2 percent in 1973. This represents an enormous increase in nuclear power production. Between 1976 and 1990 primary energy inputs for electricity generation are expected to rise from 1000 mtoe to nearly 2000 mtoe. Over half of this incremental demand for primary fuels will have to be met by nuclear power. Thus by 1990, additional nuclear capacity must be in place so that nuclear power can supply energy equivalent to over one-third of the current oil production of OPEC countries or over four times the peak oil production anticipated from the Alaskan North Slope or almost three times the peak oil production we expect from the North Sea.

Such massive increases in nuclear energy production are vitally necessary if our societies are to attain their economic and social goals. Yet unless governments implement strong nuclear programs, including international cooperation action to resolve difficulties over nuclear supplies and the common problems in the back end of the nuclear fuel cycle, much of this incremental nuclear production will be at risk. And there simply is no alternative energy as readily available to replace any nuclear contribution that we may lose, with the result that there will be a direct impact on economic growth.

If we are to overcome the current range of problems associated with expanded development of nuclear power, we must recognize that national solutions will not be enough. At all stages of the nuclear fuel cycle, the basic problems are international in character and require international solutions. This means that although specific energy policy actions will to a large extent be implemented in a national context, a high degree of international coordination will be necessary. In some cases—for example, regional fuel-reprocessing centers—a complete international solution may be the only realistic answer to the complex array of technical, economic, political, and security issues involved.

Let us review each stage of the fuel cycle to assess the international aspects. First, a comprehensive and internationally acceptable assessment of world uranium resources is required in order to determine whether an adequate reserve exists to fuel the expected growth in nuclear power and to help with judgments as to the possible timing of large-scale introduction of newer reactor technology. And because uranium in large quantities is likely to remain concentrated among a number of key producing areas, stable international supply arrangements are also necessary to allow reliable planning of nuclear power production in uranium-importing nations. In addition, international exchange of basic nuclear technology will be required if countries without a nuclear industry of their own are to have sufficient access to nuclear energy supplies.

Second is the question of fuel cycle services. Uranium enrichment, fuel reprocessing, and storage and disposal of used fuel elements and waste all involve international trade in these technologies. Reliable conditions governing the supply of adequate fuel cycle services are essential if nuclear power is to become a credible option to many countries.

Third, how can we develop a secure basis for nuclear power expansion and avoid the dangers of nuclear proliferation? International cooperation is at the heart of the solution, and until internationally accepted arrangements are agreed upon, trade in nuclear materials and equipment will continue to be subject to political and security constraints.

The main constraints to nuclear power development—including the problem of public acceptance, difficulties over uranium supply, restrictions on availability of fuel cycle services, and the dangers of nuclear proliferation—all demand an international solution. Finally, the fact that the energy problem itself is an international one needs to be understood: nations cannot make judgments on nuclear power without affecting the total supply and demand for other resources. Perhaps such understanding will help illuminate the public debate on nuclear power.

Just twenty-five years ago, nuclear power offered mankind a hopeful vision of the future—unlimited supplies of a new form of energy to take over from oil just as oil had taken over from coal. Instead, today we find ourselves confronted with a paralyzing crisis of confidence in the future of nuclear power at a time when we can least afford it. Only through international cooperation on the key issues involved can we ensure that our original hopes for nuclear power will become reality.

NOTE

1. *The Energy Challenge of the 1980's* (Paris: IEA, 1977).

Chapter 2

Uranium Supply and Demand

*Terence Price**

Since the middle of the 1970s, the physical and political availability of uranium have been central to the international political debate on nuclear power policy and nonproliferation. The Ford-Mitre study,[1] carried out by an influential U.S. team and published in April 1977 implied: that there would be no problem over uranium supply; that once the demand was there the supply would grow as needed—(as has been the case for most commodities); that there would be not any immediate need for fuel reprocessing and recycling, nor for breeding technology; and no need to separate plutonium in a pure form—at any rate not for a long time. This argument received strong support from the Carter administration; and soon afterwards, on a U.S. initiative, the International Nuclear Fuel Cycle Evaluation (INFCE) was set up at government level, in effect to demonstrate to the world that previous notions of the likely development of nuclear power could be substantially modified without harm to national energy policies.

This view was far from universally accepted. The result was that the two and a half years of work, with the cooperation of sixty-six nations and five intergovernmental organizations, that led up to the INFCE report in February 1980 was a period of intensive review of the issues

*Secretary General, the Uranium Institute, London

surrounding nuclear power, the associated fuel cycles, the raw material supply position, and nonproliferation policy. Complete consensus was not obtained, but the evaluation was undoubtedly of value in leading to agreement on a wide range of issues. Even where agreement was not forthcoming, it made explicit differences of view that might not otherwise have been brought to the surface. If it altered the balance between the 1977 U.S. views and those held elsewhere, it was probably to lend support to the position that it would still be necessary in future to follow the classical route of nuclear development—including at least some reprocessing and plutonium production and the development of breeder technology. This chapter sets out the position as seen from an industrial, rather than a political, standpoint. The views expressed are personal and commit no one but the author.

URANIUM DEMAND

Any estimate of the likely future level of nuclear fuel demand involves predicting electrical power growth, which is closely correlated with general economic growth. As countries become wealthier, the convenience of electricity attracts more uses. The result has been, in the United Kingdom, that an average 2.5 percent per annum growth in the economy (in real terms) in the decade to 1974 was accompanied by a 5 percent per annum growth in electrical production. For the world as a whole the figures were 5 and 7 percent, respectively.

The energy crisis that began in 1973–1974 and that has continued ever since has, however, had a number of consequences that make prediction more than usually hazardous. Its immediate effect was to persuade electricity authorities that their nuclear programs should be brought forward. That phase soon passed, however, when it was realized that economic activity would be permanently slowed by higher energy costs. Moreover, the long lead times for the installation of electric-power-generating capacity meant that the program planned for, say, 1975, was one that had been approved in the mid-1960s, when a quite different scenario for economic growth had been envisaged. The result was that for some time after the first phase of the energy crisis, electricity undertakings found themselves with too much, rather than too little, generating capacity. The result, inevitably, was a steady stream of cancellations, not only of nuclear plants, but also in some cases of fossil fuel stations. This trend was reinforced by conservation measures forced upon the public by higher energy costs.

With nuclear being by now a cheaper producer of electrical energy than coal and oil in many places, the nuclear industry could perhaps have looked forward to a continued swing to nuclear, in spite of the economic slowdown, once generating capacity had been brought back into line with demand. However, a number of other factors began to influence events. One was the increasing strength of the environmental lobby, which singled out nuclear power as a target and succeeded in many cases in so extending the process of approval and licensing as to shift significantly the economic balance away from nuclear, because of the capital left idle while inquiries took place. The Three Mile Island accident on 28 March 1979 induced a further pause while its lessons were being digested. So the nuclear industry entered the 1980s from a base far lower than had been predicted a few years previously. But at the same time, uncertainties regarding the political security of energy supplies continued to mount, and oil production was approaching its peak. Looking forward into the future, from early 1980 it was difficult not to believe that the long pause in nuclear development, which had by then lasted for half a decade, would soon come to an end—possibly helped by anxieties about the long-term climatic effects of continuing to burn hydrocarbons.

All this has enormously complicated the task of predicting future levels of installed nuclear power. Even without Three Mile Island and the antinuclear campaign, predictions were already extremely sensitive to assumptions regarding economic growth. If, to simplify, we take nuclear power as the preferred choice on economic grounds and ignore any replacement of fossil fuel stations, then it would be largely concerned with supplying increments in electrical generating capacity. A 1 percent change in average economic growth—say a reduction from 4 to 3 percent—would then translate into something like a 25 percent reduction in the number of power station orders or in the cumulative requirement for uranium over the next couple of decades. This is one reason why the forecasts made by OECD in 1973 and 1975 have since proved to be excessively optimistic. In fairness to the OECD officials, they were at that time working largely on data supplied by member governments that may not have been ready to admit (even to themselves) that economic growth was likely to be permanently slowed down. The most recent official estimates are, however, more in line with what the industry regards as its probable future. Some figures are given in Table 2–1.

Table 2–1 also includes some estimates for uranium requirements. During the 1980s these will depend not only on installed nuclear capacity, but also on the uranium required to feed the enrichment plants, which are essential for preparation of the fuel for all except a handful of the world's reactors. Like reactors, the enrichment stage also suffered from overprovision during the 1970s, and the tightness

Table 2–1. Installed Nuclear Capacity and Natural Uranium Requirements

Year	Worldwide Nuclear Capacity[a]	Quantity Uranium Needed (10³ metric tonnes of uranium element)[b] Annual	Cumulative from 1978	Reference
1978	105	26	26	—
1980	150	36 ± 8	97 ± 16	d
1985	255–280	57 ± 7	335 ± 50	d
	245–274	—	—	e
1990	410–530	87 ± 12	700 ± 100	d
	373–462	—	—	e
2000	850–1200	100 ± 160[c]	1300 ± 1900[c]	e,f

[a] These figures are for WOCA (world outside Communist area). In 1978 WOCA accounted for about 90 percent of the world's installed nuclear capacity.
[b] Assumes constant enrichment plant tails assay of 0.20 percent.
[c] Depends markedly on reactor strategies adopted.
[d] *The Balance of Supply and Demand 1978–1990: Report by the Uranium Institute.* (London: Mining Journal Books, Edenbridge, 1979).
[e] *Summary of Report of the International Nuclear Fuel Cycle Evaluations,* INVCE/PC/2/9 (Vienna: IAEA, 1980).
[f] *Uranium Resource, Production and Demand* (OECD, December 1979).

with which some enrichment contracts were drawn has led to some continuing overproduction of enriched material. It will take until toward the end of the 1980s for this distortion to work its way out of the system.

Sensitivities and Uncertainties

A number of factors could considerably widen the limits of uncertainty given in Table 2–1. One is the possible variation in "tails assay"—the uranium concentration in the reject stream of an enrichment plant. An enrichment plant can be operated within fairly wide limits of tails assay, and the actual choice will depend on the relative cost of uranium (high uranium cost—low tails assay) and energy (high energy cost—high tails assay). With present technology, the range of practical possibilities runs from around 0.16 percent to about 0.3 percent. The possibility of achieving in practice an average figure different from the value of 0.2 percent assumed in preparing Table 2–1 introduces an additional uncertainty into the figures for uranium requirements. For 1990 the additional uncertainty in the annual figure could amount to several thousand tonnes.

The demand for natural uranium is also sensitive to whether or not uranium—and eventually plutonium—are recycled in fuel-reprocessing plants. Recycling of uranium alone will allow uranium requirements to be reduced by 10 percent. It will be some time before enough reprocessing plant capacity will be available to permit this, but by 1990 the quantity of uranium recovered could amount in total to perhaps 40 thousand tonnes. Once fuel recycling really gets into its stride—around the end of the century—the combined effect of recycling both uranium and plutonium could reduce annual uranium requirements by perhaps 25 percent, provided, of course, that the economics of doing this are favorable in comparison with the price of newly mined uranium.

As already mentioned, both nuclear power planners and uranium producers have had to contend with unquantifiable delays, stemming from

Delays in obtaining planning permission (Australia, Canada, Japan, Sweden);
Success of nuclear objectors in invoking legal obstacles to start-up (Germany);
Delays in obtaining operating approvals owing to a mandatory need to define, for example, the exact way in which nuclear fuel is to be stored (Sweden);
Possible delays in authorization of reactor programs, owing to political uncertainties created by a case-by-case approval system applied by supplier countries (see the final section of this chapter).

However, in spite of these uncertainties, the prospects of the nuclear power industry are, by the standards of most commercial operations, still extremely encouraging. Even though they fall short of earlier expectations, a more than threefold increase in installed nuclear generating capacity between 1978 and 1990 still appears possible, with a further doubling by the year 2000.

URANIUM SUPPLY

Early History

Forecasts are most easily carried out by extrapolating well-established trends. Unfortunately, the uranium-producing industry, throughout its entire history, has never been in a state remotely approaching equilibrium. It has been either expanding rapidly or faced with near collapse. Its past history (Figure 2–1) is therefore no clear guide to what might happen in the future. Nevertheless, some understanding of that history is desirable, if only to demonstrate the vulnerability of the mining industry to factors over which it has no control.

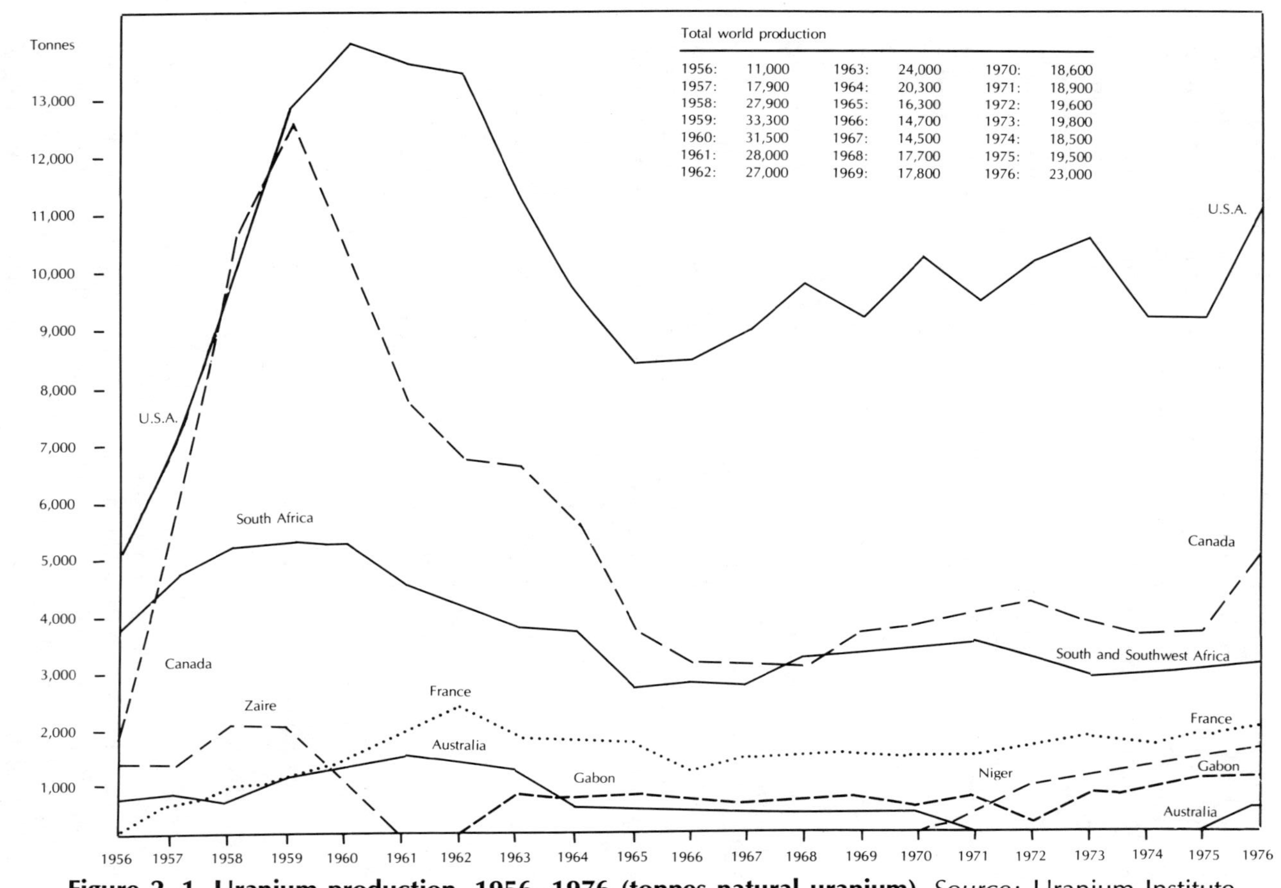

Figure 2–1. Uranium production, 1956–1976 (tonnes natural uranium). *Source:* Uranium Institute.

The commercial development of uranium began only in World War II, as a result of military requirements. At the time, the United States had not identified any indigenous resources, and its needs were met from production in Canada and Zaire. The military requirement continued after the War, and to meet it, procurement arrangements for the United States and the United Kingdom were handled by the Combined Development Agency (CDA)—a tripartite body set up by the U.S., Canadian, and U.K. governments. The CDA assisted Canada and South Africa in developing their uranium-mining production—the latter mainly as a by-product of the gold industry—and also sponsored the first of the Australian mines (Rum Jungle). The industry experienced a boom until the second half of the 1950s, and by 1957 the expansion had exceeded expectations, the incentives having proved too successful and the military requirements also having changed. The U.S. Atomic Energy Commission and the United Kingdom Atomic Energy Agency attempted to cut back on purchases from all sources, and at the end of 1959, options on contracts with Canadian producers were not renewed. As a further measure, the United States passed the Private Ownership of Special Nuclear Materials Act of 24 August 1964, which included an embargo on imports of uranium. The embargo was later lifted in stages, phased over the period 1977–1983.

In the second half of the 1960s there was a marked revival of confidence in the future of nuclear power. As a result, a number of mines were opened, some important discoveries were made (for example, Agnew Lake, Canada, 1965), and exploration expanded (in the United States to about three times the previous peak of 1957). It proved, however, that this was a false dawn, as commissioning dates of nuclear power plants were set back by construction delays and by the beginnings of a much closer governmental interest in reactor safety and licensing. The industry had to retrench once more. The embargo remained in force. Australia went out of production altogether in 1971 and did not begin again—initially on a small scale—until 1976. Exploration also declined (Figure 2–2) and recommenced in earnest only in 1975.

One incidental consequence of the collapsed markets of the 1960s was that producers were often forced to mine preferentially the highest grade ores, which were cheapest to produce. Some of the lower grade ores were permanently bypassed in the process. The policy also had the effect of distorting the productivity statistics of the industry, which subsequently gave rise to some overoptimistic and erroneous estimates of future uranium production capabilities. When production continued, it was often on the basis of long-term contracts, let in some cases at marginal prices, that later had to be renegotiated in order to allow the

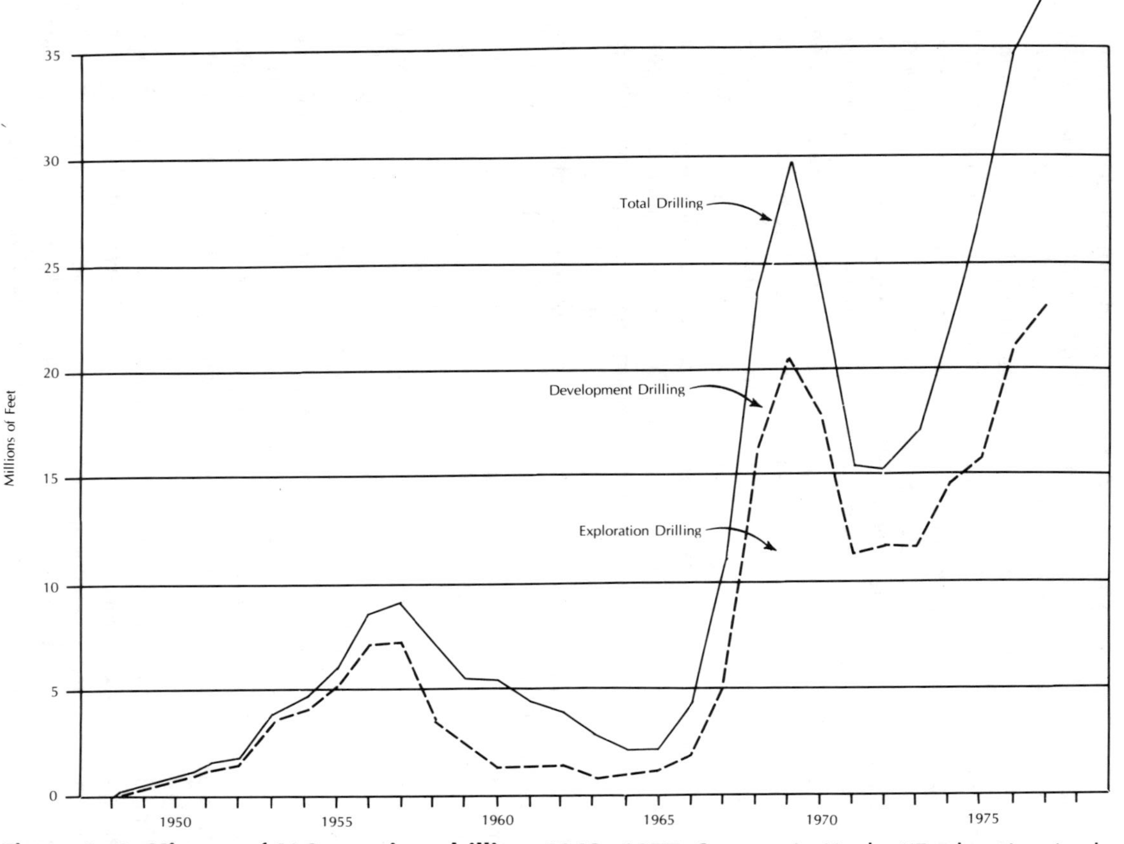

Figure 2–2. History of U.S. uranium drilling, 1948–1977. *Source:* L. Koch, "Exploration in the United States of America" (paper presented at a meeting of the Atomic Industrial Forum, Geneva, September 1976).

industry to survive in today's inflationary world and to generate the income to finance the rapid increase in production that was required.

Although the outlook for nuclear power improved somewhat in the early 1970s, recovery was slow until the whole energy scene changed dramatically as a result of the events following the 1973 Arab–Israeli war. There was a scramble for the relatively small amounts of uranium that were not already committed on long-term contracts, and the price for immediate delivery rose very sharply. The massive inflation that followed the energy crisis reinforced this trend. So, it is widely believed in the industry (though the former United States Energy Research and Development Administration disputed this), did a change in the contractual terms under which the U.S. undertook uranium enrichment contracts for the worldwide electrical supply industry. Prior to December 1972 enrichment was provided on the basis of "requirement contracts," which allowed the customer to give as little as two years notice of the date when enrichment material would be needed. This created some operational difficulties for the enrichment plants. Starting in September 1973 new enrichment contracts were negotiated on the basis of a "fixed commitment" for delivery and thus also for shipping of the feed uranium by the customer to the plant. Utilities were required to commit themselves to firm ten-year programs; and as there were substantial penalties for delay or cancellation, they were under increased pressure to ensure that their future supplies of uranium were safeguarded.

The change in the terms for enrichment services came almost simultaneously with the energy crisis of October 1973, and the effects of the two in increasing demand for uranium (both act in the same direction) were therefore difficult to disentangle. But the combined effect of increased demand and, later, of inflation was that the price of uranium oxide in the small marginal or exchange market continued to rise very sharply during 1974 and 1975. The rise may well have been spurred on by the revelation that the Westinghouse Corporation, who were large traders in uranium, had gone short of the commodity on the almost unheard of scale of 1.3 times the total annual production of the non-Communist world. The litigation and counterlitigation arising from this event has proved to be one further disturbing factor in the already checkered history of uranium.

MEDIUM-TERM SUPPLY

Uranium is a fairly abundant element that occurs in the earth's crust in an average concentration of two to three parts per million—a figure that implies a worldwide total in excess of 10^{13} tonnes. This figure is, however, of only academic interest, because what matters for practical economic purposes is the amount that can be discovered and mined within the limits of cost that the market can bear. This focuses attention on the higher grade ores—with concentrations ranging upwards from around 0.1 percent to in some cases very much higher values. The large Denison mine in Canada averages 0.87 percent, the Australian Jabiluka deposit 0.3 percent; while one of the smaller ore bodies at Cluff Lake, Saskatchewan, averages as much as 7 percent U_3O_8. Lower concentrations not only add to the problems of mining, but also create environmental problems because of the volume of rock that has to be mined and crushed in order to extract the product. Nevertheless, projects have been considered for the exploitation of ores with concentrations as low as 300 parts per million (for example, Ranstad, Sweden).

The present pattern of production (Table 2–2) is not primarily a matter of geology, but is more a reflection of the Western world's

Table 2–2. Uranium Production, 1977–1990 (1000 tonnes U/year)[a]

	1977[b]	*1980*	*1985*	*1990*
Australia	0.36	0.5	11.8–13.6	13.6–16.5
Brazil	—	0.2	0.5–1	1–2
Canada	5.79	7.3–7.6	9.2–12.5	12–15
France	2.24	2.6–3	3–3.5	3.5–4.3
Gabon	0.91	1	1.2–1.5	1.4–1.5
Mexico	—	0.2	0.6–1	0.8–2
Niger	1.44	3.3–4	6–7.5	8–11
South Africa and Namibia	5.74	10.8–11.5	11.5–12.5	12–16
Spain	0.19	0.4–0.8	0.8–1.2	0.8–1.2
United States	11.46	17.5–19	18–22	24–35
Other[c]	0.43	0.8–1.2	1.5–2.5	1.5–3
Totals				
Annual	28.6	44–49	64–79	78–108
Cumulative		120–125	395–460	760–940

[a] For the period beyond 1980, two figures are given for each country. The first figure is an estimate of realistic production capability; the second indicates the effective maximum production capability that could be achieved under favorable conditions, based on currently known resources. The countries listed in this table are those that may be expected to achieve a maximum production capability of at least 1000 tonnes U/year by 1990.

[b] The data given for 1977 are actual production figures.

[c] Argentina, Central African Empire, Federal Republic of Germany, India, Italy, Japan, Philippines, Portugal, Sweden, Turkey, Yugoslavia.

Source: The Balance of Supply and Demand 1978–1990: Report by the Uranium Institute (London: Mining Journal Books, Edenbridge, 1979).

reaction to the military demand of the 1950s, already mentioned. It is only since 1973 that the incentive has existed to look intensively for uranium in the developing world. It is to be expected that by the end of the century, uranium production will be much more widely based. But it does not follow that all the principal users will, in time, be able to rely on their own indigenous production. There will remain wide variations in the extent to which user countries can expect to be self-sufficient. The United Kingdom and Japan are both likely to remain dependent on imports.

To meet the current growth in demand, a major expansion is now under way in the mining industry. A comparison of Tables 2–1 and 2–2 shows that production should be capable, in principal, of keeping pace with demand without difficulty until at least the early 1990s. In practice, however, some qualifications to this statement need to be borne in mind.

First, uranium mining is nowadays as susceptible to planning delays as the electrical power industry, and it is difficult to anticipate the extent to which regulatory procedures will slow down expansion in future. But there will be strong economic incentives for producer governments not to prolong inquiries unnecessarily. Second, production plans are largely in the hands of private industry, which will make its own judgment of down-side risks and uncertainties in demand. We can expect a climate of indecision on the nuclear power side to be reflected in reductions in plans for uranium production. Third, uranium no longer follows the normal chemical rule of one atom being indistinguishable from another: safeguards of various kinds now attach to some uranium, though not to all, and unless some more universal system arises in time as an outcome of INFCE, this could conceivably influence the availability and distribution of the uranium that is actually produced.

Precise prediction of the world potential for uranium production beyond 1990 is impossible at this stage, but some impressions can be obtained by looking at the various contributory factors—resource availability, exploration, cost, and government policies. These factors must be judged in the light of a demand pattern that, depending on the assumptions, implies that production needs to grow at something like 7 percent per annum compound for many years until eventually breeding brings down the demand. For comparison, copper grew by a factor of 3.5 in thirty years (to 1916) or 4.2 percent per year. Oil grew 4.3 times between 1950 and 1970—which is 7.5 percent per annum—but oil exploration was able to use more powerful tools than are yet available for uranium.

It is impossible to say how far the better technology available today will make such a growth rate a practical possibility. Certainly with deposits that can be surface mined, very large-scale operations are possible. But a great deal of uranium will need to come from underground mining, where the physical limitations of ore bodies—size, shape, and depth—will impose constraints on the mining rates.

There is also the effect of falling ore grade: work tends to start off on the best ore bodies and then to progress into the more difficult areas or areas of lower concentration. This means that progressively more rock has to be mined for a given level of production, so that output falls unless more milling equipment is installed. In the United States, uranium concentrate production remained almost constant at 12–13,000 short tons of oxide between 1968 and 1976, in spite of a 50 percent increase in ore-processing rates, almost entirely because the average ore grade had fallen in the meantime from 0.21 percent to below 0.16 percent.

Further, the supply of underground mining labor cannot be taken for granted. As the labor force expands, its experience tends to be diluted; and a reduced level of skill can contribute to a poorer average return, equivalent in effect to working with a lower ore grade. While none of this is an absolute bar to attaining the required production rates, it does serve as a warning not to assume that satisfactory overall figures for resource availability will lead easily and automatically to the production of "yellow cake" on the required scale.

Resource Availability

Apart from the rate of growth, the absolute size of reserves is a further major factor. World reserves are estimated periodically by the IAEA, working in conjunction with the Nuclear Energy Agency of the OECD, in terms of the likely cost of exploiting particular deposits. The cost enters into the assessment because minerals occur in nature in varying concentrations, with the low concentration ores being more costly to mine. An increase in selling price in effect converts useless mineralization into exploitable ore and thus adds to the economically exploitable resources.

The process of discovery proceeds step by step, starting with rough general indications and continuing until sufficient is known about a deposit for firm production decisions to be made. Methods of reporting the size of ore bodies must allow for this spectrum of uncertainty. A variety of terms has been in current use, but to simplify reporting, the International Atomic Energy Agency uses only two categories for each band of costs. The term "reasonably assured resources" refers to uranium that occurs in known mineral deposits, of such size, grade, and configuration that it could be recovered within the given production cost ranges with currently proven mining and processing technology. The reasonably assured resources shown in Table 2–3 at below $80/Kg U can be regarded as roughly equivalent to "reserves" in the traditional mining sense. "Estimated additional resources" refer to additional uranium surmised to occur in unexplored extensions of known deposits (or in undiscovered deposits of known uranium districts) that is expected to be discoverable and that could be produced in the given cost range. It does not include any uranium districts which have yet to be discovered.

Figures published in 1980 as part of the INFCE studies[2] show total world resources as around 5 million tonnes U, at costs of up to $130 per kilogram of contained uranium (Table 2–3). Such resources would cover world requirements up to about the year 2010—the date depending of course on the exact assumptions. However, if we focus on the lifetime requirements of reactors in place, then it is clear from a simple multiplication of the annual requirements figures in Table 2–1 by the design lifetime of thirty years that by the mid-1990s, fears could arise about the ability of the mining industry to satisfy demand later on, unless a substantial addition to the resources had occurred by then. Furthermore, processing to recycle plutonium and unused uranium gives only a few years' grace, because of the exponential nature of the predicted growth.

Table 2–3. Uranium Resources (WOCA, January 1979; 1000 metric tonnes contained uranium)[a]

	Resonably assured resources		*Estimated additional resources*	
	Up to $80/kg U	*$80 to $130/kg U*	*Up to $80/kg U*	*$80 to $130/kg U*
	RESERVES*			
North America	752	224	1145	759
Africa	609	167	139	124
Australia	290	9	47	6
Europe	66	325	49	49
Asia	40	6	1	23
South America	97	5	99	6
WOCA total (rounded)	1850	740	1480	970

*The first column of figures roughly corresponds to normal mining reserves.

Source: Summary of Report of the International Nuclear Fuel Cycle Evaluation, INFCE/PC/2/9 (Vienna: IAEA, 1980).

Thus in order to make nuclear fuel policy hang together, something else is needed—either a slowing down of the nuclear program; or the introduction of breeding reactors, with their capacity for multiplying the energy production from a given quantity of uranium by a factor of up to 50; or an assurance that exploration will keep pace with the expanding requirement. Of these three possibilities, the first seems increasingly unlikely. The plain fact is that nuclear power is one of the very few ways in which world energy demands can be met at a time when hydrocarbon fuels will be coming under increasing strain. The second possibility is relevant to the worldwide debate on the need for reprocessing—for without reprocessing the necessary plutonium is not available, and the option is closed. The third—exploration—is a matter for the uranium industry and is clearly crucial.

Exploration

The OECD/IAEA conventions for reporting the probable size of world uranium resources could mislead a casual reader in one important respect. About half the total figure refers not to uranium that has been precisely located, but to inferences made about the probable occurrence of uranium, given what is already known about other deposits. This means that industrial exploration discoveries may or may not represent additions to the world total; that depends on whether or not their existence has or has not already been "surmised."

From the moment that a deposit is located to the time when it can be commercially exploited takes in practice something like a decade. The time taken is spent in:

Drilling out the ore body to define its grade and size;
Development to support the mining plan and engineering design;
Environmental studies;
Preparation of the environmental impact statement;
Design of mining and milling facilities;
Governmental approvals unrelated to environmental considerations (e.g., export policy);
Production financing;
Construction, including construction of infrastructure; and
Commissioning and start-up.

This process is closely parallel to what has to be undertaken for nuclear power stations and for somewhat similar reasons. As with power stations, the time spent in obtaining government approval can introduce major and often unpredictable delays. The consequent slippage in uranium production schedules is currently a significant factor in world supply, and the industry is not necessarily in a position to use delays on the power station side to prepare itself for future production.

Ideally, exploration should progress so that the world's known total resources do not diminish—which means that discoveries in a given year should match anticipated production several years later. The industry's target is for exploration to lead, if possible, by a decade. There are some difficulties in ascertaining whether this target is being achieved. Exploration results are not always announced. Drilling activity is more often publicized, but the wide differences in local geology greatly complicate attempts to infer the likely yield from the total drilling program.

There is of course always the possibility of discovering further relatively large deposits, such as the Alligator Rivers ore bodies in Australia or the ore bodies found in recent years in Saskatchewan. A few such discoveries could make a substantial improvement in the outlook. Nevertheless, when weighing future options, it must not be forgotten that exploration is essentially a chancy business and that there is still no firm evidence that it will be possible to keep pace with the progressively higher requirements of the twenty-first century. Moreover, a ten-year forward criterion is not the most stringent that could be chosen. If we use an alternative test of discovering in one year the amount of uranium needed for the lifetime requirements of reactors commissioned in that year, then the task appears considerably more demanding.

Exploration prospects will, of course, alter as time passes. One obviously helpful factor is that more experience will be built up. Developments in theoretical geology and experience in combining a variety of exploration techniques should help to improve the return. In addition, only a small proportion of the earth's landmass has yet been thoroughly explored. Even if we discard the areas that are at present of minor interest (on geological, political, or logistic grounds) or that are already devoted to other noncompatible uses (such as urban and agricultural land), we are left with about one-quarter of the world's land surface. This is comparable with the area that has already been covered, which implies that a useful factor may still be in reserve—though not necessarily at the same cost, because of the poorer accessibility that would often apply.

We may also look forward to a steady improvement in exploration instrumentation. The early discoveries of the 1950s were made largely with the aid of radiation detectors, which located surface outcrops of radioactive minerals. Low-flying aircraft provided what was in effect a mass production approach to shortlisting sites suitable for follow-up ground survey. Unfortunately, such direct methods are limited by the relatively small range of nuclear radiations in the earth—not much more than a few meters. However, a number of indirect methods are available. In some cases, surface detectors relying on radon emissions can make use of seepage through fissures to extend the range of detection down to a few hundred feet below the surface. Lower depths must be attacked by an integrated approach based on all available methods—including, for instance, inferences from theoretical geology or geochemical analysis of stream sediment or morainic boulders, supported by fence drilling in the final target area.

While nothing comparable with the "magic" potential of seismology for oil exploration seems to be in sight, particularly at the lower depths that will have to be explored as ore bodies nearer the surface are worked out, experience is constantly being accumulated. There are no reasons for doubting that the mining industry will be able to provide reliable supplies of nuclear fuel, at least until the breeder reactor is developed to a point at which we can benefit from its enormously improved efficiency for extracting energy from uranium.

Conclusions Regarding the Ability of the Uranium Mining Industry to Meet Demand

In the medium term, until at least the mid-1990s, there should be no worries about the industry's ability to satisfy demand, even if very stringent tests are applied. Exploration will certainly add to the reserves, thus stretching out the safe period. In the longer term, however, we cannot yet be certain that there will be no problems. A scenario in which nuclear power generation grows at the rate implied in Table 2–1 is one that is likely to be followed by a continuing growth in demand, at least until base load generation requirements are dealt with. Hopefully, living standards around the world will still be rising, which would mean continuing growth in electricity consumption. Population will also be going up, although admittedly this influence will be concentrated mainly in the less developed countries. The combination of these various factors takes us beyond the point where we can say with confidence how long into the twenty-first century uranium will be sufficiently abundant to make it unnecessary to place major reliance on the fast breeder.

Even if we assume that the fast breeder will be introduced progressively after about 1995, it is still not possible at the present time to give more than the roughest estimates of its effect on uranium demand. A great deal depends on the breeding doubling time. Present-day breeder reactors tend to have doubling times of over twenty years (that is, an annual gain of around 3 percent). Clearly, even if economic growth stays at quite modest levels, such breeders will take a very long time to catch up with demand; and even if breeders with considerably shorter doubling times become available, as seems not impossible, the annual requirement for fissile material will continue to grow for many years. The best that can be hoped for is for the annual demand to peak around the year 2025 and to fall to a relatively low level after the middle of the next century. By then the cumulative requirement for natural uranium will have reached something of the order of 15–20 million tonnes, even with breeders, and possibly a good deal more. It is the impossibility of being certain, with our present knowledge, of being able to find and mine uranium on this scale that drives the nuclear industry to insist on the need for breeding. Without breeding, at least to European eyes, nuclear power would be merely a transient phenomenon.

Thus, in a very real sense the breeder is a potential benefit to the uranium producers—even though it produces more fissile material than it consumes—because it enables the electricity industry to regard nuclear power as a permanent feature of the future and therefore worth pushing hard. The FBR is likely to be more costly than thermal reactors, by an amount that seems likely to support a doubling or even a trebling of the uranium price—which at present price levels is not yet a dominant element in the total cost of nuclear electricity production. The significance of this observation is that there are some theoretical grounds[3] for believing that a trebling of the price could lead to something like a fifteenfold increase in the amount of uranium in the earth's crust that would then be economically mineable. Provided always that it can be discovered, there is here at least the possibility of an extra supply cushion on which the world's power industry might be able to call. Nevertheless, the FBR is likely to maintain its significance as the ultimate insurance policy against possible future difficulties in finding, winning, or trading uranium.

POLITICAL ASPECTS OF URANIUM SUPPLY

Quite apart from these technical and economic considerations, another problem—the political availability of uranium—began to exercise the industry during the 1970s. The 1973 oil crisis served notice on energy users that massive economic disruption could be caused to states that did not control their own sources of energy. This was certainly one of the factors that gave fresh support to nuclear energy programs in 1974 and 1975—when the assumption that because the major uranium-producing countries were politically stable, the security of nuclear fuel supplies could be taken for granted. Subsequent events have somewhat shaken this view in ways that have proved to hold particular significance for nonproliferation policymaking.

The anxieties stem from several sources. First, by no means all the developed countries have their own uranium supplies. The United States, Canada, and France are well-provided; Japan, Germany, and the United Kingdom are not. Countries in the latter group understand that they cannot count on uranium imports on a scale that would deprive exporting countries like Canada of the uranium needed to fuel their own nuclear power programs. Canada has, in fact, made her position quite clear in this respect, which is helpful to other countries in making their plans.

In other respects the recipient nations have had more cause to be worried. The second half of the 1970s witnessed delays in uranium production and embargoes on uranium deliveries arising from a variety of causes. One source of delay has been in obtaining planning permission to commence mining. Two well-known cases are the Fox inquiry into mining in the Australian Northern Territory, which led to almost a three year holdup, and the Bayda inquiry in Saskatchewan. It is fortunate that these delays occurred before demand for uranium had started to grow rapidly.

The export interruptions that occurred in the late 1970s were felt by the industry to be potentially more serious in their long-term implications, even though the underlying reasons had the sympathy and understanding of almost all countries. In the case of Canadian uranium, the case was a reaction to the disturbance to local public opinion caused by India's "peaceful nuclear explosion" of May 1974. The plutonium for the explosion had been derived from reactor hardware supplied by Canada. The Canadian federal government, on 20 December 1974, announced a strengthened safeguards policy. Later on this policy—for reasons connected with an unwillingness to open the way to discretionary exceptions—became entangled with the provision of the Euratom Treaty that (in theory at least) permits free movement of fissile material within the Euratom countries, two of which are nuclear weapons states. As a result European states, which regard themselves as politically stable and reliable and which had every reason to believe that others so regarded them, found themselves for a time embargoed from receiving deliveries of Canadian uranium. There were somewhat parallel problems in connection with future Australian deliveries to Europe. There were also difficulties over U.S. enriched uranium in the early days of the Carter administration.

It would be an exaggeration to suggest that these incidents caused much more than administrative inconvenience, though at least one European utility was forced to look for alternative sources of uranium to meet the feed requirement for a stringently worded enrichment contract. Their real significance is that they served notice on all the countries affected that in the absence of binding international agreements, no one could count on the supply of nuclear fuel remaining uninterrupted, whatever the source. As a result, many voices were heard, particularly in Europe, supporting the need to complete the downstream nuclear fuel cycle capability with full uranium and plutonium reprocessing, to maintain the priority of breeder development, and to encourage the diversification of supplies of enriched uranium—all of which messages are in opposition to that of the Ford-Mitre report.[1]

Such developments will, however, take many years; and meanwhile, countries that have embarked on major nuclear power programs must continue to live with the requirements of the supplier states—unless, like France, they have their own indigenous uranium supplies. The anxieties of the consumer countries are still only partly resolved. In particular they are wary of case-by-case rules whereby the United States, for instance, can exercise control over the future disposal of used nuclear fuel in such a way as to constrain the freedom of European operators of nuclear-fuel-reprocessing plants—like those at Windscale and Cap La Hague—from handling foreign fuel. This seems to Europeans to be not only difficult to accept in political terms, but also technically unjustified. Additionally, the unpredictability of case-by-case controls, as they have been operated—and as set out, for instance, in the U.S. Non-Proliferation Act of 1978, which is a document of quite remarkable complexity—is generally regarded as a substantial constraint on an industry that, more than almost any other, is one where long lead times are unavoidable when preparing forward plans. (The full benefit of the fast reactor, on which work started in earnest in 1950, will not be realized until about 2040.)

Fortunately, both the electricity industry and the uranium-mining industries are fairly sophisticated. Both live close to government, for several obvious reasons, and they understand the need for firm nonproliferation policies. Both sides of the industry contributed to the INFCE exercise, in an attempt to assist governments to find some modus vivendi that would combine good nonproliferation controls with a regime in which trade in nuclear raw materials could be carried on in future with fewer uncertainties. The main need is for the rules governing controls on uranium trade to be codified, so that even if formal case-by-case decisions have to continue indefinitely, for political reasons, the outcome will be predictable except in the most unusual circumstances.

It would be neatest to have a single internationally acceptable set of norms, setting out the rules under which governments are prepared to allow the uranium market to operate. A single set of rules may not be negotiable, however; nor is one absolutely necessary, provided a broad consensus can be drawn out of the INFCE deliberations. That would be in everyone's long-term interests. At the very least, it would damp down the present tendency for each major country to go it alone, which can hardly assist international plans to deal with nonproliferation. It should also help to create a healthy nuclear industry, operating without anxieties about the long-term adequacy of natural uranium fuel supplies—over which there can be no final certainty for many years to come. The industry believes that, in arriving at an international consensus, the FBR and recycling must be realistically dealt with as essential components of the fuel cycle. It sees no future in policies that attempt to hold back technologies for which there are clear long-term requirements. It is to be hoped that the immense efforts that went into the INFCE exercise will assist the convergence of views and practices that will be needed if nuclear power is to take its natural place in the spectrum of energy production without giving rise to constant fears of nuclear proliferation.

NOTES

1. *Nuclear Power Issues and Choices: Report of the Nuclear Energy Policy Study Group* (Cambridge, Mass.: Ballinger Publishing Company, 1977).
2. Summary of Report of the International Nuclear Fuel Cycle Evaluation, INFCE/PC/2/9 also Report of Working Group 1 (INFCE/PC/2/1) (Vienna: International Atomic Energy Agency, 1980).
3. K. S. Deffeyes and I. D. MacGregor, "World Uranium Sources," *Scientific American* (January 1980).

Chapter 3

Conditions of World Nuclear Trade

*Ross Campbell**

INTRODUCTION

I do not propose to review in detail the changes that have come about in the conditions for nuclear trade over the past three decades, but would like to make one or two general observations. The first is that the changes that have taken place have resulted from an awareness by a number of states that changes were needed to keep pace with advancing technology and with wider experience of its uses and—more importantly—potential abuses. Some, though not all, require a political as distinct from either an institutional or a technical solution. A second general observation is that the concept of a world made up of supplier states and recipient states is giving way to one that acknowledges that a dialogue among members of one group in the absence of the other will not produce enduring results. I shall turn to these thoughts later.

Before looking at the conditions for nuclear trade that might be obtained over the next decade or two we should quickly review the current conditions under which such trade takes place. It must be recognized that with only a few exceptions, most suppliers are be-

*Chairman, Atomic Energy of Canada Ltd.

holden to others for some part of their nuclear needs, and hence the distinction between nuclear haves and have nots is not as clear as one might think. Although we have a group identified as nuclear suppliers consisting of some fifteen states, on examination, most members of the group will be found to be dependent to some extent on other members of the group or on others outside the group for some of their nuclear needs. In short, nuclear trade is characterized by a very high degree of interdependence.

CURRENT CONDITIONS OF NUCLEAR TRADE

Limitations under which nuclear trade today must be conducted include conditions:

1. The majority of states engaged in nuclear utilization or trade are parties to the Treaty on the Non-Proliferation of Nuclear Weapons, and although the parties do not unanimously agree that all of the terms are being fulfilled, there is nevertheless agreement that identified nuclear transfers into nonnuclear weapon states will be covered by safeguards administered by the IAEA. Parenthetically, I might mention here that Canada as a nonnuclear weapon state party to the treaty has its entire nuclear program under IAEA safeguards, and hence our imports of nuclear material and equipment—and we do have some—are subject to treaty safeguards.

A number of parties to the treaty recognized that some of its terms would require further interpretation in order to put treaty language into practice, and this resulted in the so-called Nuclear Suppliers Guidelines that were promulgated by the IAEA in February 1978. While I agree that there was a need to reach a more definitive agreement as to the conditions under which nuclear trade might proceed, it is my personal belief that nuclear commerce might have been better served if the discussion and subsequent arrangements had included more states and if a more determined effort had been made to reach absolute unity of commitment among the fifteen. The reservations entered by some states, including my own, robbed the guidelines of some of their force.

2. Trade in nuclear equipment and materials has been and will likely continue to be a high profile activity that engages the attention of a small but nevertheless vocal segment of the population of many states. Under the stimulus of such groups, the desirability of using nuclear power to produce electricity is being called into question in many countries and gaining sufficient popular attention to cause the

desirability of exporting the means to produce such power to be similarly challenged.

3. There is agreement among a significant number of states that electrical energy needs will have to be met in growing measure by nuclear power. Of these, a number feel that energy independence presupposes not only the use of nuclear power but the ability as well to develop to the extent possible an indigenous capability to design and build nuclear power stations. To make sales, some nuclear plant suppliers are entering into agreements to transfer technology to customers that in years to come may themselves become suppliers.

4. A preoccupation with the management of nuclear waste is one that is common to virtually all states engaged in nuclear activities and is perhaps paramount today among the concerns that have been identified in public forums. Matters that have been given public prominence throughout the history of the development of nuclear power have included reactor safety, physical protection of nuclear materials, proliferation of nuclear weapons, and waste management. Although it cannot be claimed that all but waste management are no longer issues, it seems to me that a satisfactory demonstration that wastes can be managed effectively is a matter that deserves priority attention not simply because antinuclear groups have concentrated on it as a weak link in the chain but rather because it is incumbent on those who have developed this timely new form of energy to be able to provide it free of any suggestion of bequeathing a difficult legacy to future generations.

5. Nuclear power is regarded today as a proven energy producer by most financial institutions, and the early reluctance to finance the construction of nuclear power stations has disappeared. Although such stations are capital intensive and although they must compete for funds with other large projects, there does not appear to be any slowdown in trade through lack of loan funds from private and government sources. There is, however, a need for a more forthcoming attitude on the part of international lending institutions in meeting the nuclear energy needs of the developing countries that lack alternative means of power generation.

6. The almost global slowdown in nuclear power installation that has arisen is the result of other factors that are at play. One of the paradoxes of our time is that there is general agreement that alternative energy sources are needed to either augment or replace those that are based on fossil fuels, yet one alternative, nuclear power, with a proven capability to respond to that need safely, reliably, and economically is being denied its full potential.

7. Nuclear trade in the context of this discussion is carried out under the umbrella of either bilateral or multilateral agreements or both and

therefore involves participation by governments. The governments of the principals in the transaction are likely to be closely associated with this trade, while other governments will follow it to the extent that interest dictates and diplomacy allows. Given governmental interest in, and preoccupation with, nuclear trade, sales tend to be regarded as national sales as distinct from corporate sales and therefore have additional criteria that must be met.

8. Nuclear trade today involves a number of suppliers with whom buyers may negotiate, and the market today with some exceptions may be characterized as a buyers' market. Their preferred position notwithstanding, buyers feel in many cases that if they are to proceed with the deployment of nuclear power plants, they will need to have greater confidence in continuing supply and other associated arrangements than exists at present.

IMPROVED CONDITIONS OF NUCLEAR TRADE

It is against this background then that one must try to identify routes whereby conditions for world trade may become more favorable than is the case today. Clearly there are a number of favorable conditions, and these we need to retain. Much of the ground that has been won by nuclear proponents, suppliers and users alike, cannot and should not be lost. To give but one example, a number of states, my own included, have mounted and maintained an impressive national effort to bring a nuclear industry into being. Those that were involved in such programs will vividly recall trying to get funds for civilian nuclear power development during the period when oil was priced below $2 a barrel and thought to be in inexhaustible supply. These deserve credit for persuasiveness and tenacity in having carried the day, and I think that I would not be guilty of condescension if I were to suggest that the governments of the day deserve their fair share of the credit.

In Canada, successive Canadian governments lent support to AECL during periods when the power development phase of the program was under attack. Happily, support was continued, and thus a power system was in place, proven and ready to respond appropriately to the challenge precipitated by the events of September 1973. We were not unique in this respect, and all nations similarly placed must ensure that the technology remains alive and well so that it may respond to future needs, both national and international. An incalculable disservice would be done if the current slowdown in nuclear plant orders results in the dismantling or the deterioration of segments of the nuclear industry, with the result that when it is needed—and I say when, not if—it will not be in a position to respond. Unless conditions permitting the more widespread utilization of nuclear power as an energy source can be created, we will be courting an energy crisis in the 1990s that could have far-reaching social and economic consequences, not excluding tensions among nations.

The foregoing is not exhaustive, but I think that I have identified a sufficient number of conditions affecting nuclear utilization to illustrate the atmosphere in which nuclear trade is operating today and, more particularly, the severe restraints surrounding that trade, some or all of which are susceptible to amelioration. What can be done to improve that atmosphere, so that tomorrow's nuclear energy needs may be met in a timely, ordered, and businesslike fashion?

I would like to start with some views as to what might be done in the commercial arena and thereafter address technical, institutional, presentational, and political adjustments that in my view need to be made if nuclear power is to reach its full potential in the decades ahead. In the space available, it is not possible to discuss any one of these areas in depth—let alone all four. Accordingly, I will submit for consideration in each area a suggestion that could contribute toward an improved atmosphere for world nuclear trade.

COMMERCIAL

I wonder if I am too far off the mark in suggesting that if a way could be found to establish what I call the credibility of contracts, a major commercial concern of both buyers and sellers alike would be removed. Uncertainty as to whether a contract will be fulfilled in accordance with its terms is not solely a concern of buyers—it is simply that sellers' concerns tend to be less publicized. In the commercial nuclear world there has been and continues to be a mutual concern that events beyond the control of both parties may interfere with the completion of a contract. I am not talking here about force majeure events, for clearly, trading partners have coped with such events in the past and will be able to do so in the future. I am talking about events occasioned by decisions taken by national authorities that can affect contracting parties in diverse ways, and I will return to this point later on when discussing safeguards.

TECHNICAL

In identifying waste management as the technical area that should be addressed, I am conscious of the fact that some will take the position that it is not a technical problem at all—or at least not an intractable one—but rather a political or perhaps a social problem. In large measure I agree. It seems apparent nevertheless that further demonstration of waste management techniques is needed at both the national and the international level. This is not to suggest that a significant amount of work is not going on at both levels, and I am heartened by the cooperative effort in waste management studies among states and international agencies alike. I would, however, like to see governments take a more positive public stand on waste management matters than has been the case to date. I am personally persuaded after discussions with my own technical people and with those in other countries that the technology and infrastructure associated with the safe storage and management of nuclear wastes is not only meeting existing needs but will be ready to meet future needs as well.

But the fact that I am personally persuaded does not of course contribute very much to the formation of opinion among the general public whose support is needed if nuclear power is to achieve its full potential as an energy source. Because of nuclear power's historical associations with military use and because it is based on a science little understood by the public, efforts to address the issues publicly face an emotional barrier. Antinuclear groups capitalize on this emotional aspect by attempting to cast doubt on the credibility of those associated with the nuclear industry, including their highly reputable scientists and engineers. This is not to suggest that those of us in the industry should stop trying; on the contrary, if we do our public information job as well as we do our technical job, it will be a further step in the right direction. But more is needed—especially a more positive public stance by persons elected to all levels of government. A recent joint announcement by the Canadian minister of energy, mines, and resources and the Ontario provincial minister of energy launching a joint federal-provincial nuclear waste management development program provided visible government support for the Canadian waste management program. I am convinced that if our politicians would continue to speak out in support of both nuclear power and waste management programs, they would find an audience that those of us in the industry have been unable to reach.

Such support by governments is all the more needed to prevent articulate minorities from bypassing and usurping the function of the democratically elected organs of society. Protest and pressure groups, initially hailed as participatory democracy in action, now threaten, through skilful manipulation of public meetings and media and through recourse to civil disobedience, the authority of institutions that nourished them. Overt government support for the concept verification stage of current waste disposal options would lend a welcome aura of legitimacy to industry efforts.

INSTITUTIONAL

There are a variety of institutional arrangements presently in place and making a significant contribution to the growth and development of nuclear trade, and I am not about to suggest that yet another international organization is needed. I think we have quite enough already. Nor am I about to suggest any reduction in either the number of organizations or their programs.

I do, however, want to focus attention on one international organization—the IAEA. Seminar participants are familiar with the agency and how effectively it discharges the mandate given to it by its statute. Some have been involved in a variety of ways with the agency since its formation and can be justifiably proud of its accomplishments. I think all will agree the agency deserves the continuing support of its member states. It is on the next point that I expect disagreement. It is nevertheless my view that the support evidenced in the recent budgetary and voluntary fund discussions is not in keeping with the importance to the world community of both the promotional and safeguarding activities of the agency. The contribution that these activities have made and are making to the growth and safe development of nuclear power utilization is of inestimable value and in my view deserves support beyond that which is currently accorded. The reverse side of this coin perhaps illustrates the point more dramatically—that is, few governments or companies would enjoy public support for the current level of nuclear trade if the agency's inspection arrangements were not available to ensure that safeguards undertakings were being fulfilled.

The agency is being asked to take on ever increasing promotional and safeguards responsibilities, and it seems clear that those activities must be supported in a manner commensurate with their importance to the needs of member states. I submit that these needs will not be met if current attitudes prevail. If the agency is seen to be financially troubled, and hence less effective, public attitudes toward commitments that are dependent upon a strong agency presence would at best be unpredictable.

PRESENTATIONAL

Given the safety record of nuclear power stations and their proven ability to produce reliable, economic power at a critical time in the world's energy history, the nuclear industry ought to be enjoying broad popular support. That it is, in fact, increasingly embattled may be attributed in part to the secrecy—for good reasons—in which the technology was born and in part to the failure—for no good reason—to educate the public about its peaceful applications once the mantle of secrecy was for the most part cast aside. Instead, those who would bend our societies in a more decentralized direction at lower standards of living have, under the guise of a concern for the environment, managed to capture public attention. They have skilfully played upon fears of genetic damage, of terrorist abuse and of catastrophic accident to bring this timely new form of power generation into some disrepute. Their goal, and let there be no misunderstanding on this point, is one of bringing about a total prohibition of nuclear development.

I would suggest that the resources of the IAEA be mobilized to become a world repository of expert, authoritative comment on each and every contentious allegation that has become the stock in trade of protest groups. The views of a respected international agency tend to be much more believable to the average citizen than those of national experts, who are often regarded as biased on one side of an issue or the other.

POLITICAL

Any discussion of conditions for world nuclear trade would be meaningless without reference to the political factors involved. Of all the items of international civilian commerce, trade in nuclear equipment, material, and technology attracts national and international attention to a unique degree. From the beginning of negotiations for a nuclear power reactor sale to the final commissioning of the facility, one is faced with obstacles that simply do not arise in any other trading activity of which I am aware. Among the many such obstacles, the quest for solutions to the political problems associated with nuclear trade is perhaps the most challenging. Prevention of the proliferation of weapons as a by-product of nuclear power programs is the major political problem.

One of the first endeavors designed to lay the groundwork for international control of nuclear energy—the Acheson-Lilienthal plan—was rejected by the USSR, and hence control through secrecy was continued until the early 1950s. It had by then become apparent that secrecy alone could not prevent the proliferation of nuclear weapons capabilities and that other measures were needed. The U.S. Atoms for Peace program and the creation of the International Atomic Energy

Agency were but two of the manifestations of the political measures taken to promote international cooperation in peaceful uses of atomic energy under the aegis of suitable international control. Although some unsafeguarded national civilian programs were continued, the majority of the world's nuclear activities were subject to IAEA safeguards, and it is noteworthy that during the period from the inception of the IAEA in 1957 to the coming into effect of the Treaty on the Non-Proliferation of Nuclear weapons (NPT) in 1970, there was no demonstrated proliferation of nuclear weapons from independent power programs. This is not to say that all nations refrained from developing nuclear weapons: two did, but the development resulted from dedicated weapons programs.

The coming into force of NPT, with the very broad support through ratification that it has received, provided further indication of the extent to which a majority of the world's nations are prepared to work toward the prevention of any increase in the number of states having a nuclear explosive capability. Although giant steps had been taken in the development of an international nonproliferation regime, events during 1974 caused a number of nations, including Canada, to reconsider the adequacy not only of national policies but of international nonproliferation mechanisms as well.

Among the "events" of that year, the Indian explosion was certainly the most dramatic, alerting the world as it did to the real risk of weapons proliferation. The other "event" of that year was the realization by the entire world in the wake of the 1973 oil embargo that a massive threat to oil supplies existed, that nations would have to free themselves from dependence on imported oil if they were not to live under an economic and political sword of Damocles, and that increased reliance would have to be placed, inter alia, on nuclear energy. Since uranium, like oil, was a finite resource, this led in turn to accelerated attention being given to more fuel-conserving nuclear fuel cycles. All involve reprocessing of irradiated fuel, and all involve material that, when separated, could be of weapons grade. Given the energy outlook today, one has to assume that reprocessing will be commonplace within the next two decades. The aim must therefore be to allow reprocessing under controlled conditions—not to prevent reprocessing altogether.

All of this did not suddenly become apparent in 1974, but it was certainly in that year that the developed nations began to give serious attention to prolonging the availability of the world's uranium resources by recourse to more fuel-conserving cycles. By 1977 the prospect of a large-scale recycling of fuel and large-scale deployment of uranium-plutonium- or thorium–U-233–fueled breeders was sufficiently real to cause genuine concern to be voiced about the equally large-scale traffic in weapons-useable material that this would entail. This, as I understand it, was the genesis of the Carter government's decision to propose an International Nuclear Fuel Cycle Evaluation program (INFCE), with the stated aim of searching for technical means of developing proliferation-resistant fuel cycles, but in reality to force a pause in the "proliferation" of enrichment facilities and reprocessing facilities, the two direct routes to a weapons capability. During this period, parallel studies could be undertaken not just for technical but for institutional ways of containing a potentially dangerous situation. It was with that aim that the May 1977 Western summit launched and subscribed to INFCE.

INFCE has been widely misunderstood as commercially motivated to prolong the U.S. monopoly on enrichment or to buy time to search for an effective breeder technology, but this is not so. American policy, as I understand it, reflects a genuine concern about the quantum jump in international use of and traffic in weapons-useable materials that now seems imminent. Canada shares this judgment. Indeed, if new rules for this nuclear future are not devised, there could be an almost uncontainable proliferation-prone state of affairs, with a great many nations operating their own reprocessing facilities, if not their own enrichment facilities, and engaging in international transfers of the fissile materials for breeders that are equally the basic materials for weapons.

A technical fix might be possible to frustrate terrorists from acquiring separated plutonium, but it will not be enough to stop any government that might be tempted to pursue a weapons program under the guise of recycling fuel for a power program. It would be an expensive route to take, but it is one that might hold some attractions for an unscrupulous government bent on acquiring a weapons capability and ready to pay a premium for secrecy and surprise.

It is this part of the problem that demands a political solution, not a technical one—and it is just possible that INFCE has the right composition of states to provide the forum in which a political solution can be sought. Unlike the Nuclear Suppliers Group, INFCE embraces not just the suppliers of nuclear technology, equipment, and materials but the principal consumers as well—and in addition includes among its participants most of the economically significant countries of the world. All are aware that they—and the world as a whole—are going to need increased recourse to nuclear power as the world's petroleum reserves come under a supply and cost squeeze about a decade hence. It is therefore in their common interest to promote conditions that will allow an upsurge in nuclear utilization. That upsurge will be blocked by public opposition founded on fear of weapons proliferation unless new and commonly agreed rules are adopted to govern this accelerated phase of nuclear use.

"Sanctions" is a word that has unpleasant and unsuccessful historical connotations. For example, we know that sanctions or embargoes imposed unilaterally have little chance of achieving the desired goal. If the sanctioned state is able to obtain embargoed items from others—as happened in the case of the Canadian embargo on nuclear trade with India—the embargo has little effect. But if we could have supplier guidelines backed by appropriate sanctions that were agreed to in advance by a number of states—supplier and recipient alike—we would have a situation in which a transgressor would have taken part in formulating the penalty being applied. The rules and the penalties for infraction would be self-imposed rather than imposed in the traditional sense. The present composition of INFCE lends itself to an effort in this direction, and there is no reason why every state, as it is about to acquire its first nuclear technology or equipment, should not become a member of the "Group of 40." Little by little a universally applicable code of conduct could develop and be established in a multilateral treaty. A concomitant of such a political compact would have to be acceptance of full scope safeguards by all INFCE states and acceptance of stepped up resources for IAEA inspection, so that timely warning of suspected diversion could be given.

I am persuaded by the record of unsuccessful attempts by other routes that the collective route to sane nuclear guidelines backed by agreed sanctions is the only viable route remaining. In historical perspective, once the immediate postwar effort to completely internationalize the control and management of sensitive nuclear activities had failed, the focus of control attention was thereafter directed toward the receiving state. That was the essence of the IAEA statute and of the accompanying inspection system. It is also the essence of the commitment undertaken by recipient states under the NPT—that nonproliferation of weapons would be achieved by strict control in the receiving state over transferred equipment and materials.

The explosion of a device by a nonnuclear weapon state in 1974 caused the onus to shift from controlling the receiver only to controlling the donor or supplier. Thus, first the almost clandestine London Suppliers Group of seven and later the more open fifteen-nation Nuclear Suppliers Group attempted to strengthen the nonproliferation regime by devising guidelines for suppliers as to the undertakings they must obtain from recipients or pledge to forego the business—a sort of control by denial of the materials and technology. But, like OPEC, suppliers' groups generate resentments on the part of those not making the rules—and within the supplier group, not all subscribe to exactly the same rules, which could cause a reaction on the part of non-suppliers that could destroy the existing nonproliferation regime. A further disadvantage of supplier-made rules is that despite an agreed set of guidelines representing a common denominator, there is room within the supplier group for a very unhealthy degree of "competition in safeguards," where one supplier can gain commercial advantage over another by offering slightly less onerous safeguards while still staying within the common denominator.

Finally, as we have seen all too clearly at this seminar, the attempts by some suppliers to combine agreed supplier guidelines with somewhat stricter domestic criteria for supply and to apply the result in a nondiscriminatory way can have unintended repercussions on other suppliers, while missing entirely the true target—the potential weapons state: you aim at Idi Amin and you hit your friends. Denial is a risky policy, and nondiscrimination can be a two-edged sword.

Clearly, neither NPT nor the Nuclear Suppliers Group guidelines, with their many imperfections, would be equal to the infinitely more demanding breeder phase of nuclear energy technology, which will involve increased traffic in materials having parallel energy and weapons capability. Almost by definition, it will be necessary to engage suppliers and receivers alike in the rules to govern such a potentially hazardous game, and once agreed, they should be the only rules applied without discrimination and with agreed penalties for infraction. The alternative to such a consensus approach is a sort of anarchy that is not at all pleasant to contemplate.

To return to the "commercial" theme of my remarks, without some such compact to put public opinion at its ease about advanced cycles and reactors, I foresee a very inhospitable climate in which the suppliers of reactors will have to conduct their business. The INFCE forum offers the opportunity, if only the participating nations will recognize soon enough the political rather than technical dimension of the problem.

Chapter 4

Nuclear Energy, Nuclear Exports, and the Nonproliferation of Nuclear Weapons

*Günter Hildenbrand**

THE NEED FOR NUCLEAR ENERGY

In approaching the subject of nuclear energy, nuclear exports, and the nonproliferation of nuclear weapons from the viewpoint of German nuclear supply industry, let me first make some general points that—although open to disagreement in matters of detail—are in principle accepted by the broad consensus of opinion. These general points, together with the subsequent discussion of the situation in the Federal Republic of Germany, are the basic elements for my views on the subject under discussion.

- The average yearly growth rate in the worldwide generation of electrical power for 1958 to 1976 was 7.4 percent (Figure 4–1). Even if economic growth is slowed down, a steady increase in worldwide electricity consumption—and thereby in required power plant capacity—is to be anticipated.
- For the power plant capacity, required to increase from about 1.8 million MW today to about 6.5 million MW by the end of this century, only the following sources can make any significant contribution—hydropower, coal, oil, natural gas, and nuclear en-

*Head, Kraftwerk Union/Germany

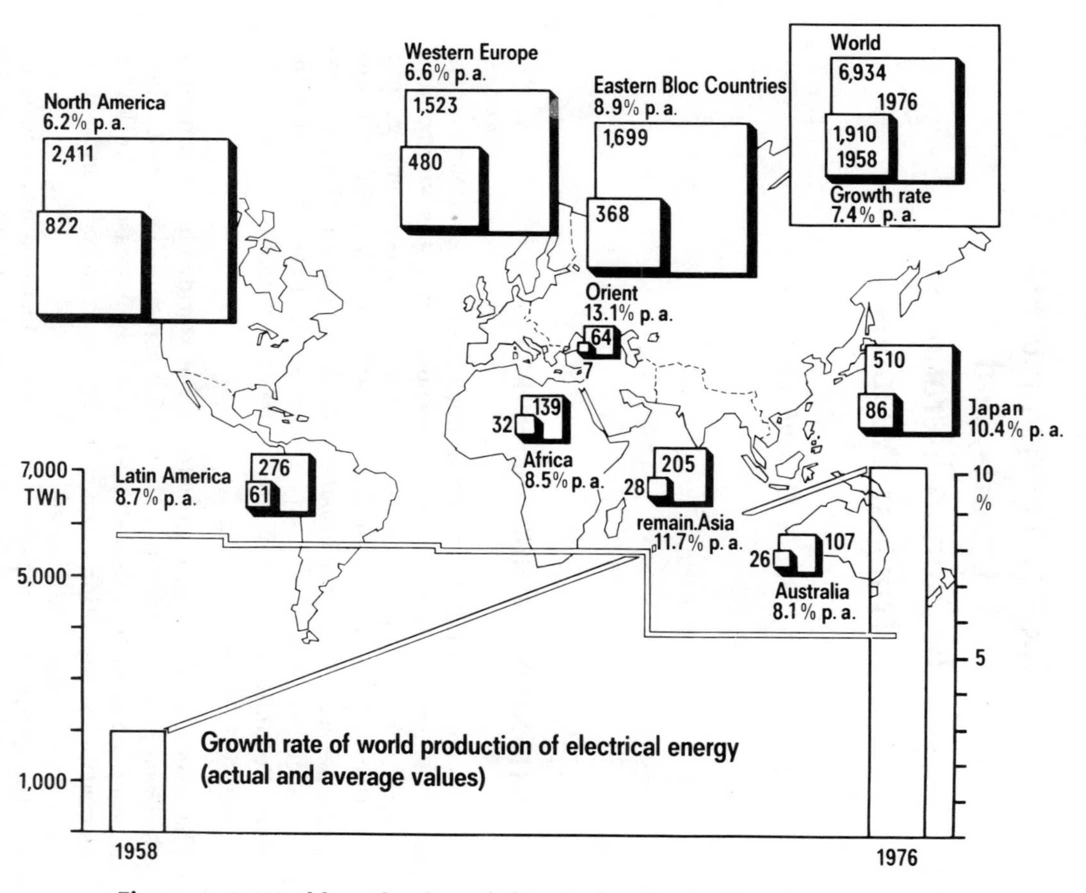

Figure 4–1. World production of electrical energy 1958 to 1976 (in TWh).

ergy. Before the end of this period, the proportion of oil and natural gas will have decreased, while that of nuclear energy will increase; coal will more and more have to be treated as a valuable raw material.

To reduce at least the relative portion of oil in the production of electrical energy is not only a question of availability in the long run but also a question of the expenses for mineral oil imports (Figure 4–2). The influence of the 1973 oil crisis can clearly be seen. In 1976 our country spent 32 billion DM to import the same quantity of oil that was imported in 1973 for 11 billion DM.

- Present opinion is that there are only two energy sources that will remain available virtually indefinitely—nuclear energy, with the option of the breeder principle (in fission and fusion reactors) and solar energy. Until the technical processes involved with fusion reactors and large-scale use of solar energy have reached full maturity, nuclear energy from water reactors and synthetic energy from the gasification of coal will become increasingly important, with nuclear energy again playing an important part as a process energy source.
- A reliable and economical supply of energy is absolutely essential for each and every country's economic development. In the interests of closing the prosperity gap between North and South, this is particularly true for the developing countries. Unrestricted access to nuclear energy, sooner or later, is therefore a perfectly legitimate aspiration for these countries.

Two facts underlie the great interest that the developing countries take in nuclear energy due to their increasing energy demand: At present the per capita energy consumption in the United States is twice as much as in the Federal Republic of Germany, whereas in many developing countries it is one-tenth of the German consumption. And second, the growth of the population in developing countries is faster than that of industrialized countries (Figure 4–3).

- Nuclear energy is not only indispensable, it is also well-suited for the needs of the future, is safe and economical, and has little impact on the environment. Elements of its orientation toward the future are the optimum fuel utilization and the use of fuel-saving reactor systems. One important element of its environmental acceptability is the safe and permanent disposal of radioactive waste.
- The need for the increased use of nuclear energy must be brought into harmony with the aims of nonproliferation.

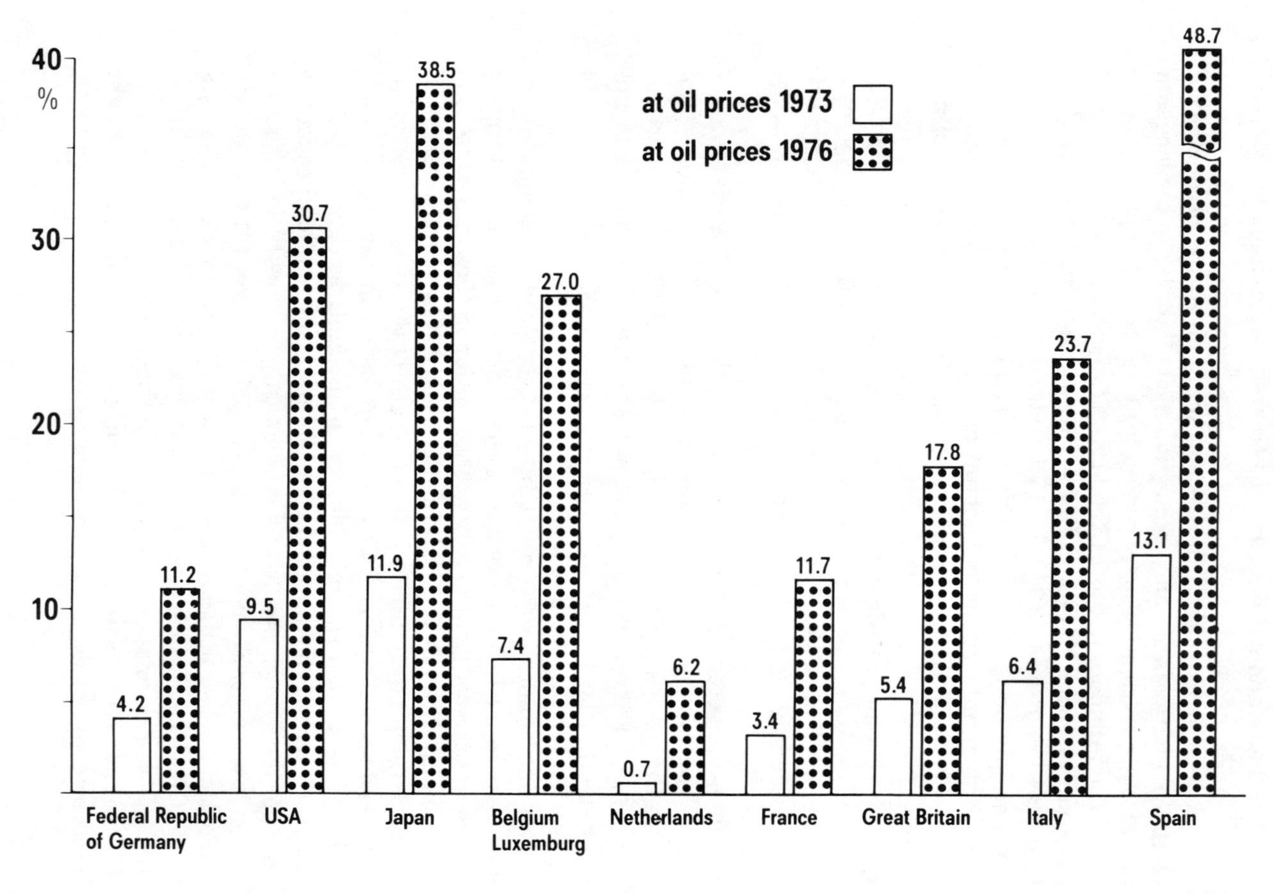

Figure 4–2. Share of the expenses for mineral oil imports in the export returns (mineral oil imports 1976 rated at oil prices of 1973 and 1976).

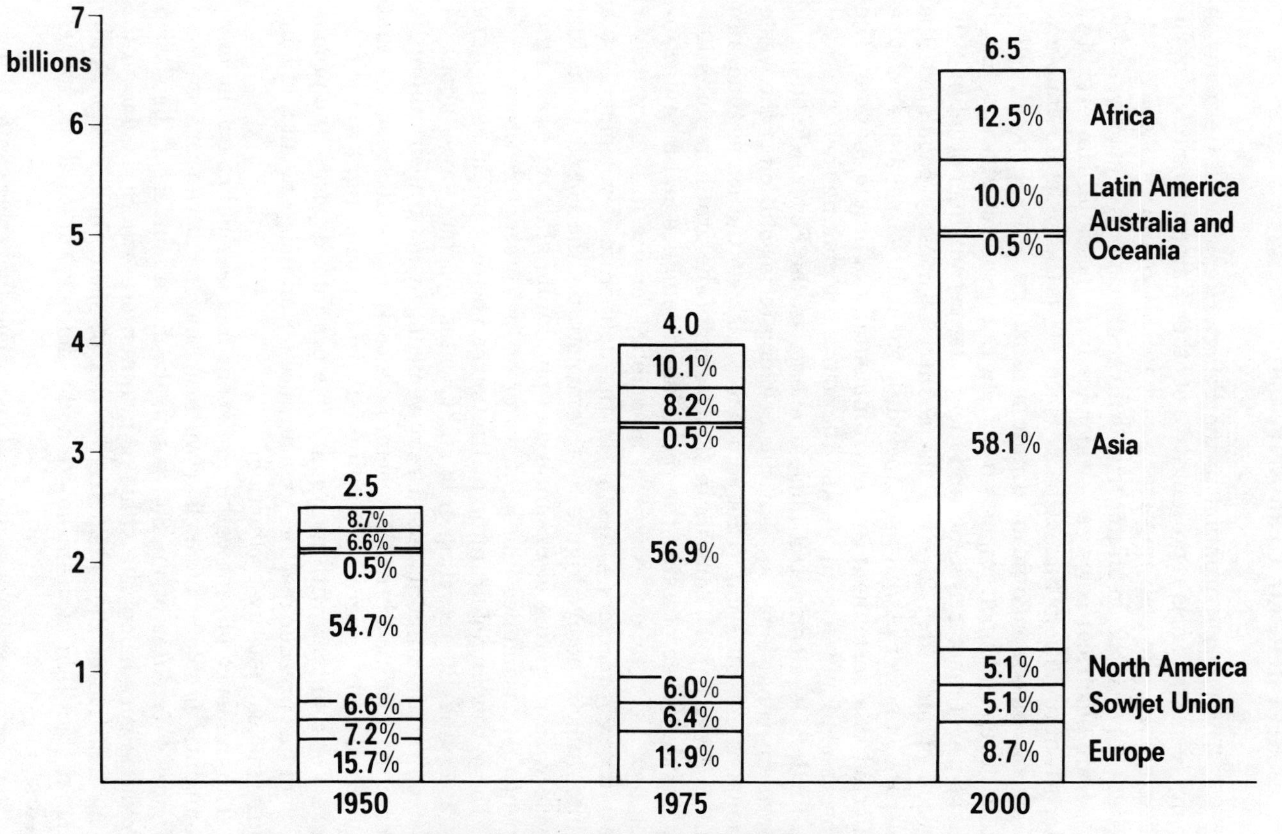

Figure 4–3. World population, subdivided to continents.

THE SITUATION IN THE FEDERAL REPUBLIC OF GERMANY

The energy situation in the Federal Republic of Germany in many respects is also representative of other European countries. Nuclear energy plays an important part in this country's endeavors to reduce dependence on oil imports, to diversify sources of supply, and to generate electrical power as economically as possible. However, two important conditions have to be satisfied: (1) as the Federal Republic of Germany has no significant uranium resources at its disposal, but is fully dependent upon imports from non-European countries, it has to make use of reactor systems and fuel cycles permitting optimum utilization of fuel; and (2) since the country is densely populated, the disposal of power plant waste, including spent fuel assemblies must be effected in a permanent and environmentally optimum manner. The government, public opinion, and the courts are making their approval of further development of nuclear energy in the Federal Republic of Germany conditional on a satisfactory long-term solution of the waste management question. This solution, which at the same time also fulfils condition (1), consists in the closing of the uranium-plutonium fuel cycle by reprocessing the spent fuel assemblies and manufacturing new U-Pu mixed oxide fuel assemblies for use in light water or fast breeder reactors—that is, in the deliberate development of a plutonium economy. Figure 4–4 demonstrates the closed U-Pu fuel cycle with recycling of uranium and plutonium after reprocessing of spent fuel assemblies into light water or fast breeder reactors.

The optimizing of uranium utilization can be seen in Tables 4–1 and 4–2. U and Pu recycling into light water reactors saves about 34 percent of the annual reload requirements in natural uranium; Pu recycling into fast breeder reactors opens the door for reactors that have a sixty times greater uranium utilization than light water reactors. For countries with no remarkable uranium deposits, the promotion of fast breeder reactors is insurance for an independent and long-term nuclear energy production.

The influence of reprocessing of spent fuel assemblies on the environmental problems of nuclear waste management can be seen from Figure 4–5. After about 600 years, the hazard index of the fission products goes below the value of uranium ore, whereas—due to the long half-live of Pu—the hazard index of spent fuel is remarkably higher than that of uranium ore and remains so up to about 100,000 years.

The German spent fuel disposal or nuclear fuel cycle center (Figure 4–6), which is at present in the project–planning stage, will not only

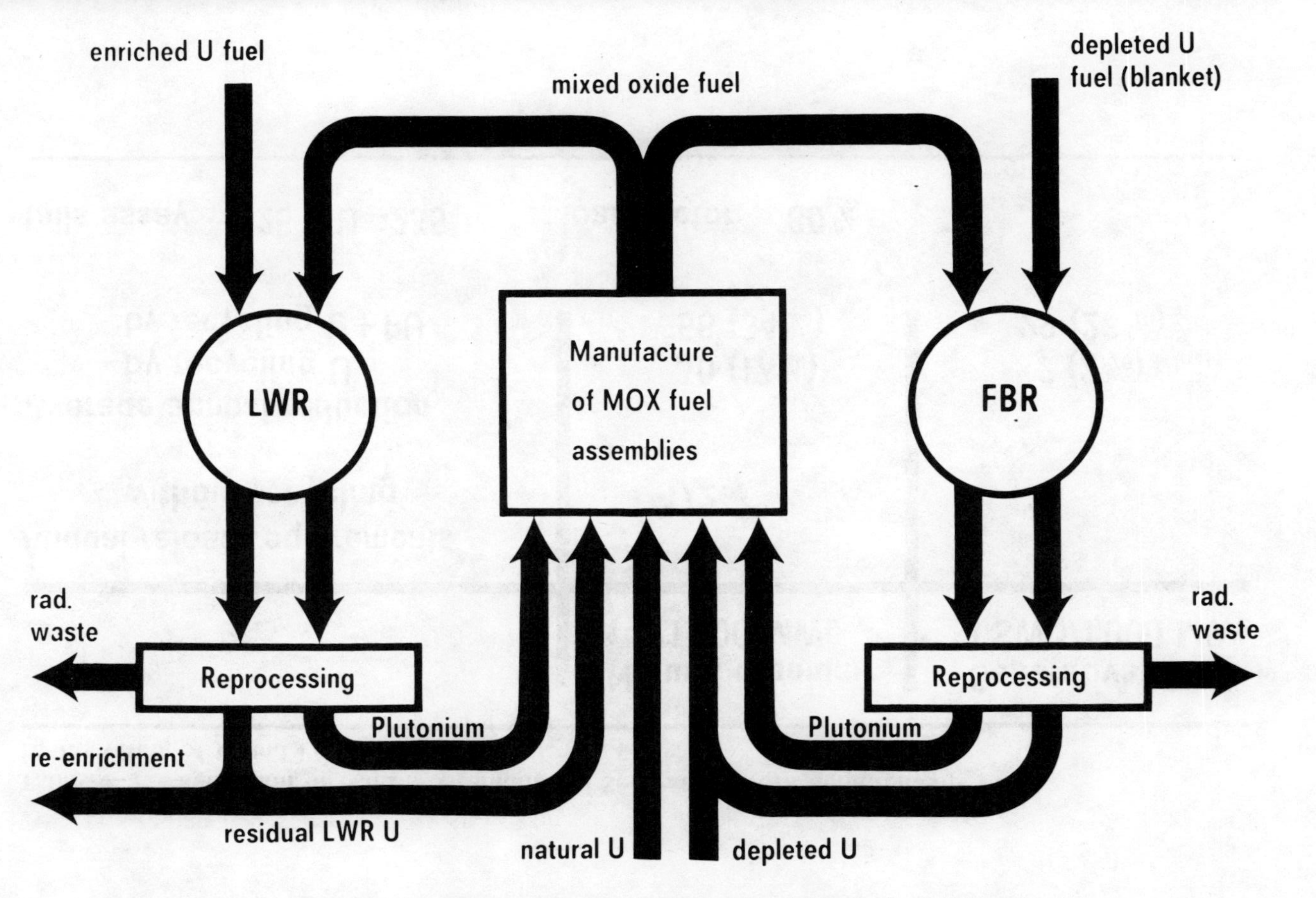

Figure 4–4. Pu recycling in LWRs and use of Pu in FBRs.

Table 4–1. Reduction of Natural Uranium and Separative Work Requirements by Recycling of U and Pu into LWRs

	Natural uranium t U/1,000 MWe	Separative work t SWU/1,000 MWe
Annual reload requirements without recycling	172	112
Average annual reduction		
by recycling U	30 (17 %)	2 (2 %)
by recycling U+PU	58 (34 %)	29 (26 %)

tails assay: 0.25 % U—235 load factor: 80 %

Table 4–2. Natural Uranium, Separative Work, and MOX Fuel Manufacturing Requirements for LWRs and FBRs

Reactor	Annual reload requirements		
	natural uranium	MOX fuel manufacturing	
	t U/1,000 MWe	kg Pu^{fiss}/1,000 MWe	t HM/1,000 MWe
LWR			
without recycling	172	—	—
with recycling (U + Pu)	114	194	6.8
FBR (oxide fuel)	3 (including blankets)	2,200	17.8 (excluding blankets)

tails assay: 0.25 % U—235   load factor: 80 %

Natural uranium and MOX fuel manufacturing requirements for LWR's + FBR's

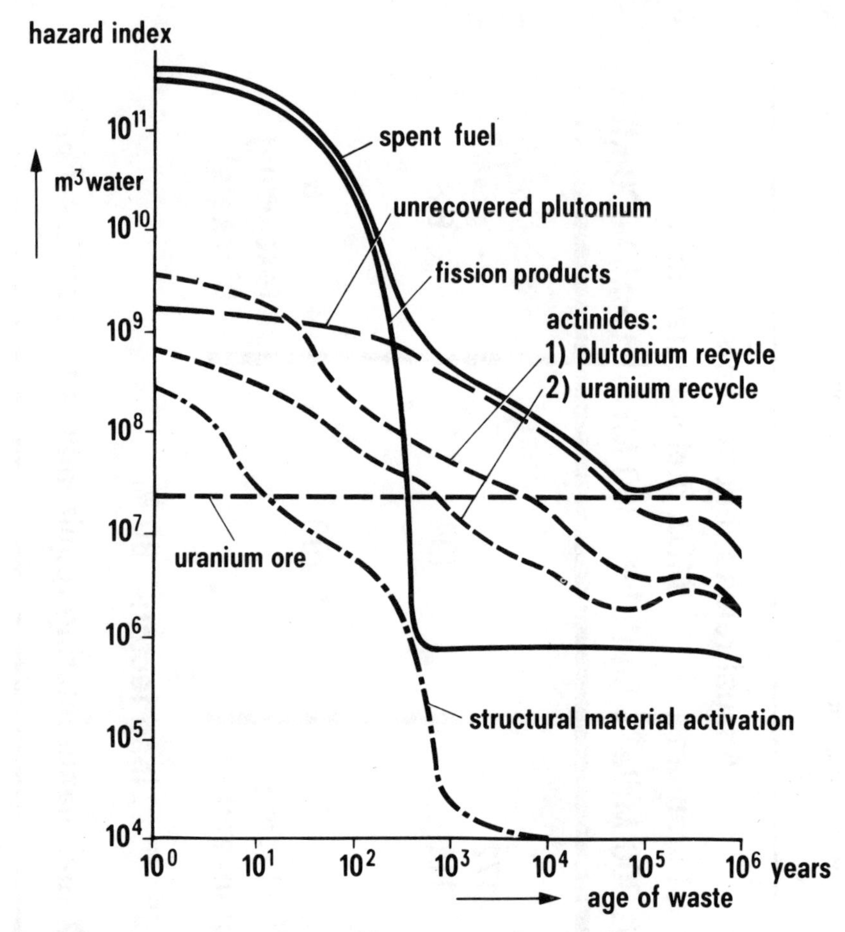

Figure 4–5. Summary of hazard index components from a 1000 MWe PWR (wastes from 1t of fuel at 34,000 MWd/t). *Source:* Bishop, Malaro, and Hewitt, *NRC and Nuclear Waste* (AED—Conf. 1977, 077–000).

fulfill the two conditions mentioned above, but will also do much to serve the aims of nonproliferation via colocation. As all systems are concentrated at one site on top of the rock salt dome for final waste deposit, the transportation of nuclear fuels along public traffic routes will be restricted to a minimum. Spent fuel assemblies come into the center and only newly manufactured mixed oxide fuel assemblies and uranium (in the form of UO_2 or UF_4) leave it again. Since the plutonium made available by reprocessing is immediately made up into new fuel assemblies, the amount of free, explosive-grade fissile

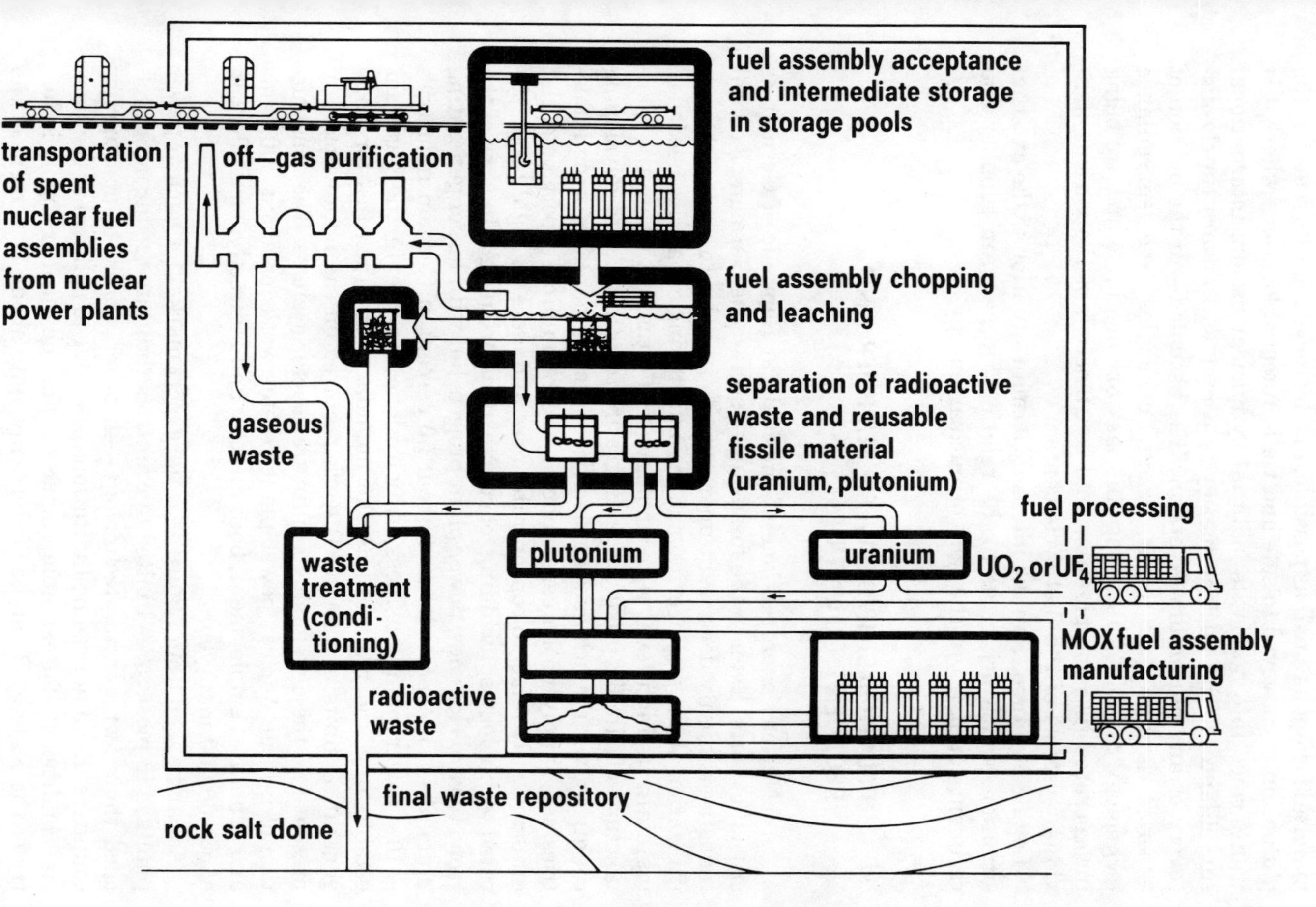

Figure 4–6. Spent fuel disposal center.

material is kept as small as possible; and by recycling into reactors, the plutonium is conveyed to safe and well-protected locations where it is transformed into nonfissile material by nuclear fission. International surveillance of the nuclear fuel cycle center will ensure timely discovery of any plutonium diversion. It is assumed—in the opinion of experts rightly—that plant design principles and surveillance technology capable of fulfilling this task are available and are being incorporated into the planning and construction of the various individual systems of the fuel cycle center.

Let me now turn from the internal German situation to the external situation, which brings us to the relationship between supplier and recipient countries in the field of nuclear energy.

RELATIONSHIP BETWEEN SUPPLIER AND RECIPIENT COUNTRIES

Nuclear energy is one of the Federal Republic of Germany's crucial exports. Even in the fossil-fueled sector, the German electrical industry is highly dependent upon exports in order to ensure the full utilization of the engineering and manufacturing capacities needed to maintain its high level of technology. In a situation where the manufacture of simple products is migrating continually to developing countries (with which as regards price German manufacturing industries can simply no longer compete), the remaining export potential automatically tends to concentrate more and more on highly sophisticated technologies, including, of course, nuclear energy. In view of the trend towards less new power plant projects but with larger generating units, it is an export share of about 50 percent of German industry's entire power plant business, which alone can create the basis for establishing and maintaining the manufacturing facilities for large generating units and for R&D work required on the improvement of present and the development of future reactor systems. The construction of a large nuclear power plant creates work for about 700 firms and generates employment, both directly and indirectly, equivalent to about 50,000 man years.

As can be seen from Figure 4–7, the export market for nuclear power plants is in most cases limited to countries that have so far not developed their own reactor technology—in other words, to nonsupplier countries. Even here some limitations exist due to political or commercial conditions—for example, financing. For the nuclear supply industry of the Federal Republic of Germany, only about 17 percent of the world market in thermal power plants is a potential market for nuclear

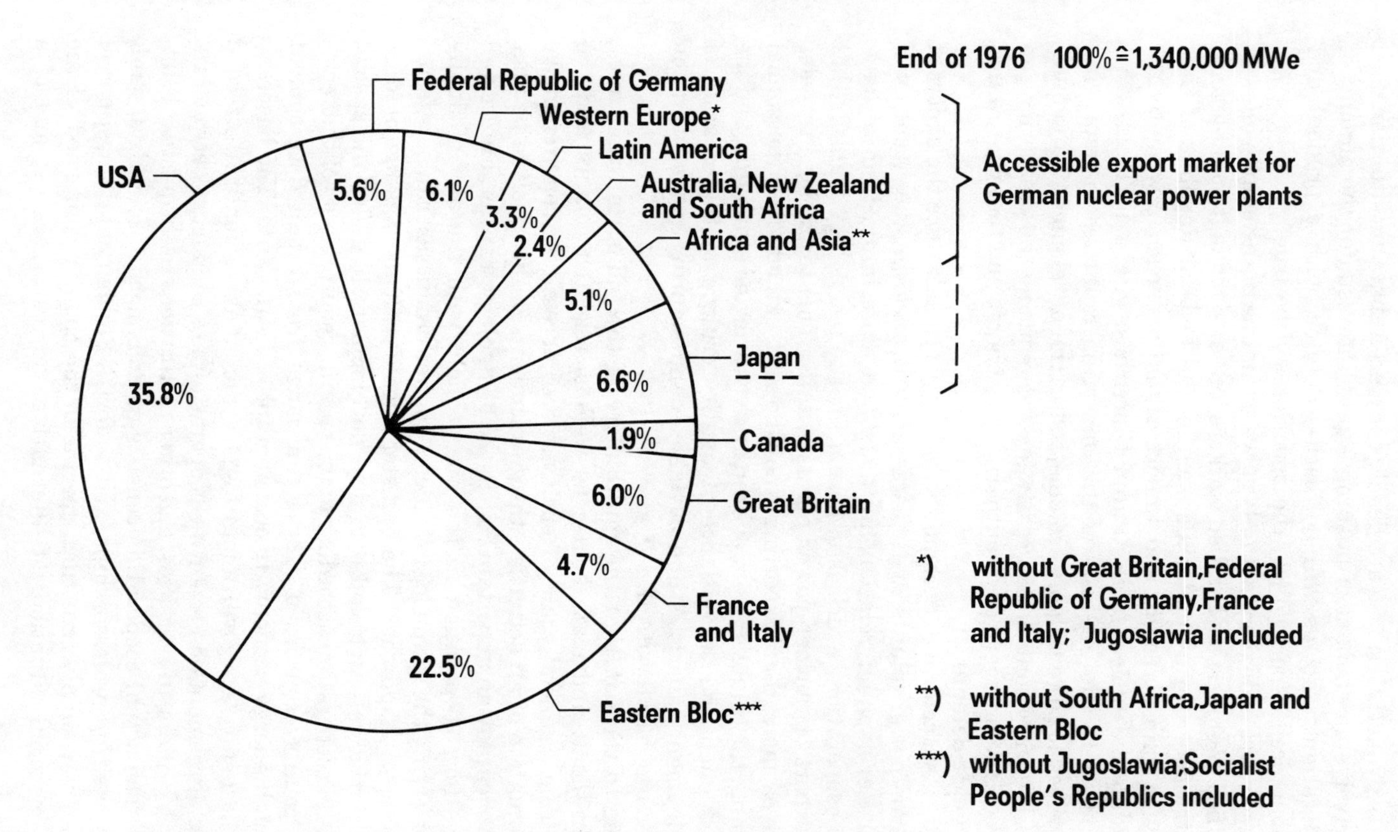

Figure 4–7. World electric power installed in thermal power plants (January 1978).

exports. Nonetheless, we are deeply convinced that this situation does not involve an element of pushing export interests for the amount of proliferation risks; as will be discussed later, with a prudent and well-balanced export policy, the opposite will be true.

Our export endeavors not only serve to maintain the employment of qualified engineers and skilled workers, but also make it possible for a country poor in raw materials such as the Federal Republic of Germany to gain access to and import urgently needed raw materials including the nuclear fuel uranium. Since these vital export endeavors coincide with the interests of the importing countries in improving their electric power supply systems, industrial development, and quality of life, they contribute greatly toward reducing the prosperity gap between the northern and southern hemispheres and thereby towards worldwide détente.

This last factor will become increasingly crucial, since the export of nuclear power plants will necessarily transform the traditional purchaser-supplier relationship more and more into a form of long-term cooperation, which is essential for the technology transfer that the export of nuclear power plants entails. This technology transfer involves more than exporting know-how; it extends from personnel training, the establishment of licensing criteria and standards, and the introduction and surveillance of quality criteria to the creation of development capacities and a viable manufacturing industry in the importing country itself.

Due to the fact that the multiplicity of the industrial companies, government, and state bodies as well as research centers and the resulting financial scale of such projects necessitate comprehensive coordination of all activities, the governments of the involved partner countries play an important role in the technology transfer. Figure 4–8 shows how cooperation in the energy field with the aim of finally transferring the technology of nuclear power production can be effected in a stepwise approach: The first stage deals with an exchange of ideas of scientific research and technical development to define the special areas of cooperation according to the specific requirements of the recipient country concerned. These can range from urban development through energy economy to oceanography, for instance. The bases for this first stage are government-to-government agreements. In a second stage agreements on scientific cooperation are conducted between research centers, universities, and other institutions in the partner countries. The actual cooperation consists in exchanges of scientists, seminars, visiting professorships, and so forth. Fundamental questions of the energy supply structure, specific to the country concerned; of the raw material situation; of alternative sources of energy; and of a

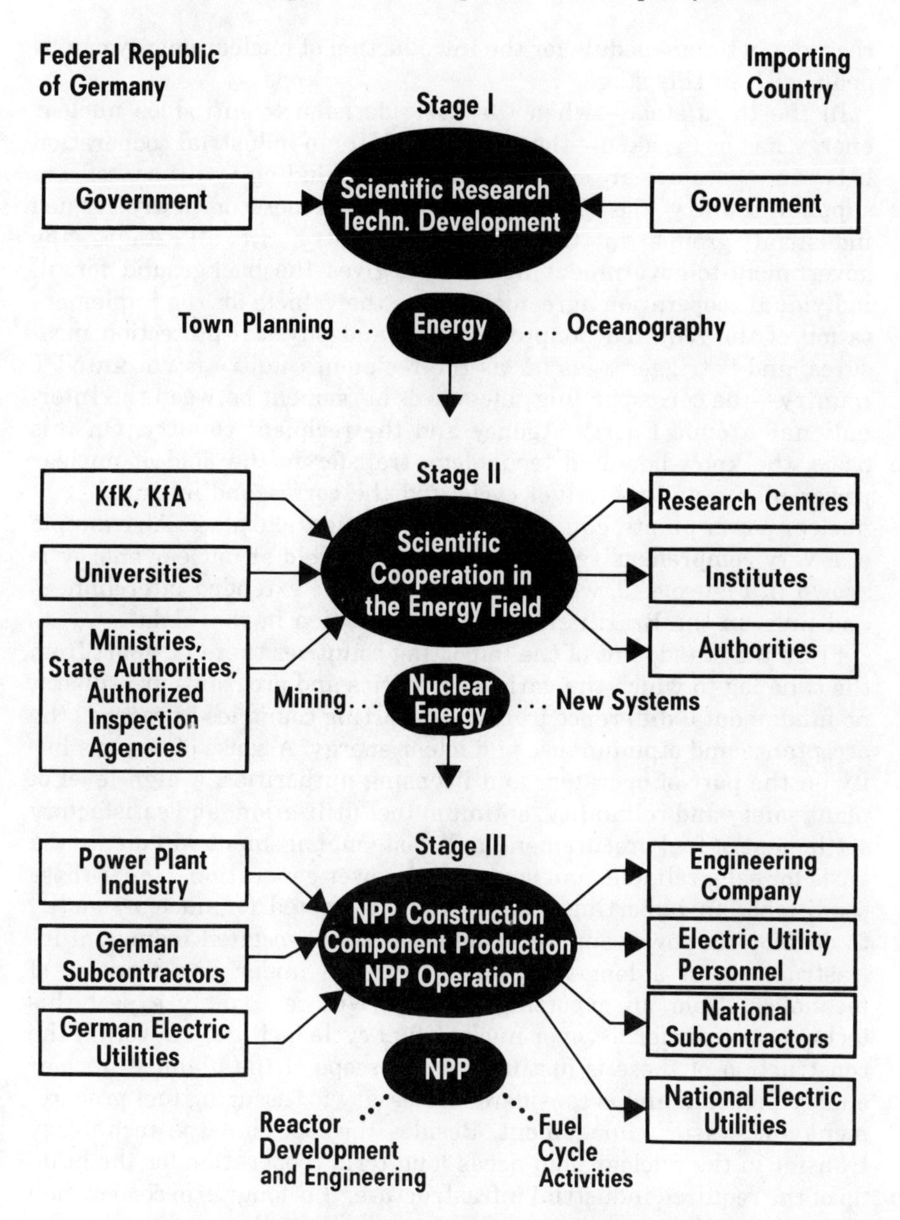

Figure 4–8. Stages in the transfer of technology.

reasonable time schedule for the introduction of nuclear energy can be dealt with in this stage.

In the third stage—when the basic decision to introduce nuclear energy has been taken—the emphasis shifts to industrial cooperation between the nuclear energy industry (including utilities) of the supplier country and existing industrial partners or newly formed industrial groups in the recipient country. In all stages the government-to-government agreement gives the background for all individual cooperation agreements; it is the vehicle for the implementation of the required nonproliferation and physical protection measures, and it triggers—as far as the recipient country is not an NPT country—the corresponding safeguards agreement between the International Atomic Energy Agency and the recipient country. On this basis, the know-how and technology transfer in the field of nuclear power plants and their fuel cycle and the corresponding supplier of nuclear power plants, equipment, and facilities take place. An example of a very comprehensive cooperation in the field of nuclear energy is shown in Figure 4–9, which demonstrates the extended interchanges and links in the Brazilian-German cooperation in that field.

From the standpoint of the importing countries there is, apart from the time lag to which the various activities and programs are subject, no fundamental difference from the exporting countries as regards the acceptance and otpimum use of nuclear energy. A sense of responsibility on the part of operators and licensing authorities, a high level of plant safety and reliability, optimum fuel utilization, and satisfactory settlement of fuel procurement and waste management will create the basis for safe, reliable, and economical power generation. Under these conditions, an importing country that has opted for nuclear energy to develop its power supply system and the associated industrial infrastructure on a long-term basis, with an underlying transfer of technology from an exporting country, will reasonably expect this technology transfer to cover nuclear fuel cycle facilities as well, if the construction of these is justified by the scope of the planned nuclear energy program and is considered necessary for assuring fuel procurement and waste management. Besides the fact that the technology transfer in the nuclear field needs long-term cooperation for the build up of the required industrial infrastructure, this long-term cooperation gives the recipient country, via corresponding license agreements, access to the results of R&D work going on in the supplier country on the improvement of existing and development of future reactor systems and facilitates its own R&D contribution where and whenever skilled manpower and financial resources are available.

Obviously, responsible exporting and importing countries will be

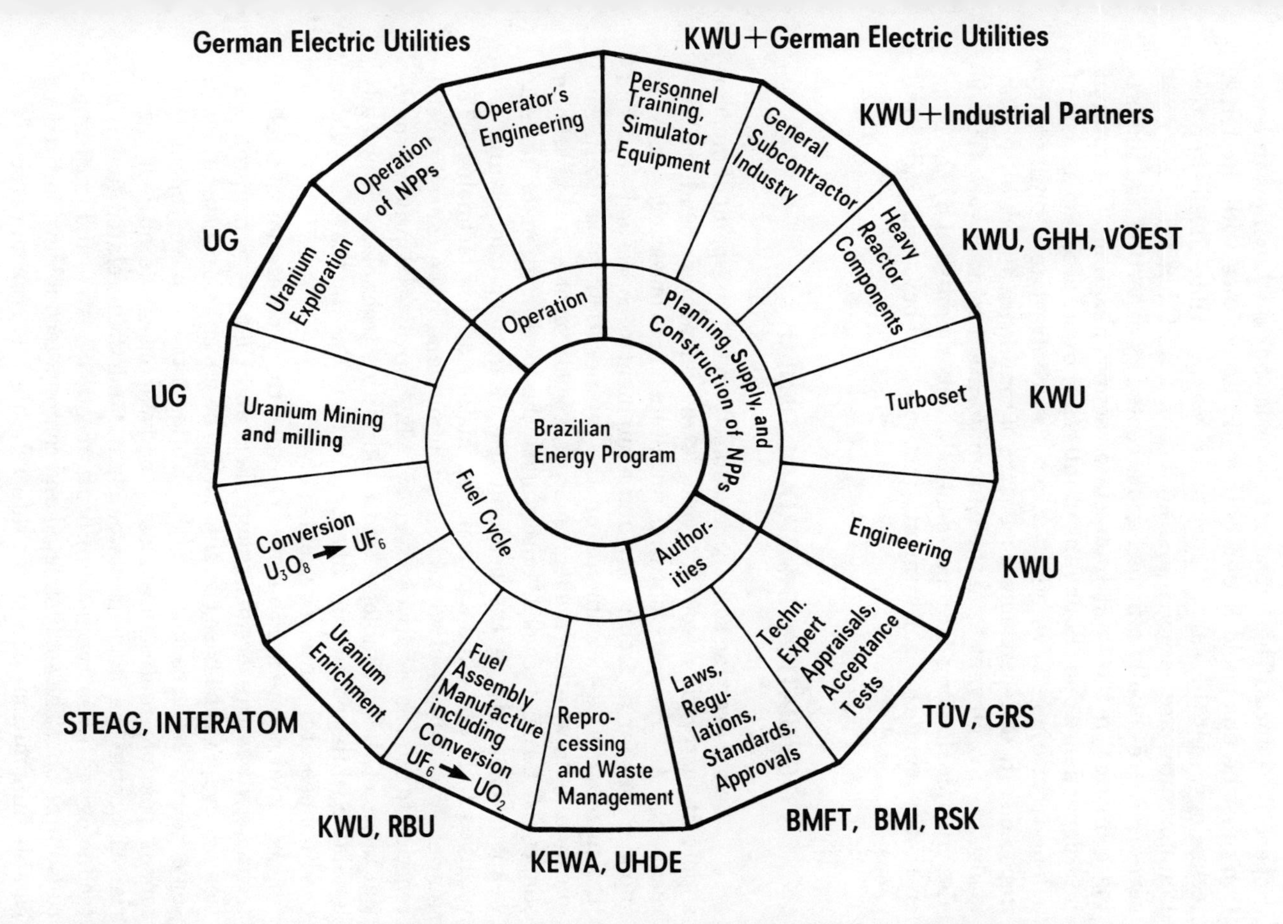

Figure 4–9. Infrastructure program of Brazilian–German cooperation in the field of nuclear energy.

equally interested in keeping to the aims of nonproliferation in their cooperative efforts. The important point is that the facilities in question must be amenable to surveillance in accordance with the latest technology, that this surveillance and the appropriate physical protection measures are accepted and maintained, and that these are measures that are equally affirmed and accepted by exporting and importing countries in the vital interests of nonproliferation. The mutual interdependence and channeling of interests, freely accepted as part of a long-term cooperative effort, creates in contrast to a unilaterally imposed state of dependence a measure of confidence and stability that is also highly relevant from the nonproliferation point of view. This is also true for the long-term personal relations established by the cooperation among the government, institute, agency, and industrial people involved.

COOPERATION INSTEAD OF DENIAL

Before drawing some conclusions from the viewpoint of nuclear supply industry in a country that is in the nuclear field both as recipient—as far as natural uranium is concerned—and as supplier—as far as technology and equipment is concerned—I would like to make a few comments on the proliferation issue itself. The question as to whether nuclear energy as such, or at least via certain shapes of its fuel cycles, encourages the proliferation of nuclear weapons is as old as the peaceful use of nuclear energy and has recurred time and time again in the history of this technology. The answers given have always been numerous and have caused official reaction to alternate between secrecy, which involves depriving potential users of this technology, and making this technology completely available while remaining fully committed to the aim that it should be used for peaceful purposes exclusively.

My central points are that, in principle, the proliferation of nuclear technology has irreversibly taken place and that every nonproliferation policy has to be based on this fact to be effective. Investigations in some countries, especially in the United States, have shown that among possible routes to explosive or weapons-grade nuclear material, the route via nuclear power plants and their related fuel cycle technology is by far the most difficult, most expensive, and most time-consuming one. The sophisticated and complex technology required for establishing nuclear power as a safe, reliable, and economical source of energy production is the reason why developing countries need our long-term cooperation. If they wanted to go the military route they could do it sooner or later and with relatively modest financial efforts by themselves. This is shown in Table 4–3, which summarizes the results of a study made in 1975–1976 by Professer Wohlstetter on the time and cost required for the development of nuclear explosives. The lead times range from one to ten years, and costs between $1 and $200 million have to be expected.

This does not exclude the argument that the use of nuclear energy for power generation could provide an umbrella under which a clandestine development of nuclear explosives could go on and could even be stimulated by the technological experience from the peaceful activities and that the threshold "to go nuclear" would thus be lowered. But on the other end stands the fact that the nuclear explosive decision is a highly political and very serious one for any country, whether it has a peaceful nuclear program or not, and that there is the unique chance by cooperating with a country embarking on the peaceful use of nuclear energy to bring one's country into the family of countries that use nuclear power for peaceful purposes exclusively. The vehicle for this is export regulations and safeguards agreements that are satisfactory according to the state of the art and are nondiscriminatory between supplier and recipient country. They should be based on a strategy of trust and cooperation in place of one of denial and maintaining dependence. They proceed from the fact that mutual interdependence must be desired and freely accepted between supplier and recipient countries if it is to be successful from the point of view of nonproliferation and that a partnership with equally balanced rights and obligations will help to make the world a safer place to live in during the nuclear age.

It could well be that sanctions would be the only answer in a specific case of proliferation. But these sanctions are based on political decisions and should not be anticipated by keeping the recipient country dependent through a strategy of denial as far as certain technologies are concerned that are, on principle, proliferated anyhow. One sometimes gets the impression that restrictions that burden the peaceful use of nuclear energy are being suggested because their inventors feel that doing something toward nonproliferation—even with the risk of acting counterproductive—would be better than having no immediate answer to the required political solutions, including sanctions. But one cannot escape the task for reasonable political actions because they would be required in any case of proliferation, whether a nonproliferation or safeguards agreement has been violated or never existed. In the latter case the warning time would certainly be less than in any case of collaboration under safeguards; in fact, it would be zero.

In view of the legitimate aspirations of all countries to cover their

Table 4–3. Time and Cost for Development of Nuclear Explosives

Category	fissile material from	minimum time (years)	cost (million $)	technological status	countries
0				nuclear weapon state	USA, USSR, UK, France, China, India
1	enrichment, reprocessing	1	0.85	existing fissile material and technology	Japan, FRG, South Africa, Belgium, Taiwan, Italy, Netherlands, Canada, Sweden
2	reprocessing enrichment	4 5	50.85 105.8	existing reactors and certain technology	Israel, Argentina, Switzerland, Egypt, Spain, South-Korea, Indonesia, GDR, Czechoslovakia, Australia, Pakistan, Iran, Norway, Brazil, Mexico
3	reactor and reprocessing	8-12	208.7	no potential	other countries

F

enrichment via gas centrifuges, reprocessing minimized to Pu extraction

(Reference: A. Wohlstetter et al., Moving Toward Life in a Nuclear Armed Crowd, 4.12.1975/22.4.1976)

ever-rising electrical energy requirements and of the fact that developing countries are much more dependent upon the assistance of the industrialized countries in developing nuclear energy for peaceful uses than they would be for military purposes, there is a great risk that a strategy of withholding know-how and maintaining dependence could bring about the opposite of what is intended—namely, political disintegration and nuclear self-sufficiency outside the worldwide nonproliferation system aimed for, to a point where certain circles could start weighing the possibilities of gaining access to nuclear weapons without infringing international agreements. Even—or precisely—those countries that have willingly accepted discrimination as nonmembers of the nuclear weapons club (by being nonweapon NPT members) because they fully support nonproliferation will regard a second discriminatory measure restricting access to and utilization of the uranium-plutonium fuel cycle technologies as a challenge to make themselves self-sufficient in nuclear fuel procurement and waste management—that is, to aim for independent national solutions. Energy policy is not only a fundamental question of economic and social stability and future development for all countries but is also a sensitive area in terms of national sovereignty. In the view of the importing countries, the risk of misuse of their dependence by an exporting country is much greater than the danger of breaking an international safeguards agreement, which in fact no one has done.

Any defense of the sovereignty of countries in dealing with their energy problems must mean that the U.S. decision to defer the setting up of a plutonium economy indefinitely must also be respected in principle by other countries. With the existing self-sufficiency in natural uranium and technology, this policy might—although difficult to understand from the outside—even be justifiable for a number of years. However, this does not exclude the right to criticism, in pointing out that the present proven nuclear technology and that of some decades to come shows the nuclear energy economy to be essentially a plutonium economy. Apart from the limited value of alternative fuel cycles for the improvement of nonproliferation, the lead times for their development up to the status of the existing uranium-plutonium fuel cycle will be in the order of twenty or even thirty years. It must therefore be the goal of the technology concerned to exploit the most highly concentrated energy source available—Namely, Pu-239, like U-235 and U-233 equivalent to about 14,000 barrels of oil per kilogram—for the generation of electrical energy while keeping to the goals of nonproliferation.

The throwaway energy policy imposed by outlawing the plutonium economy can hardly be justified, particularly in terms of the interests of other industrialized and developing countries that are dependent in the first place on imports of conventional energy sources and to an increasing degree on the nuclear fuel uranium. As far as explosive grade fissile material is concerned, it should be borne in mind, that there is no principle difference between U-233, U-235, and Pu-239 and that no inherent proliferation-proof nuclear fuel cycle exists presently or can be expected. Therefore, there is no rational justification to concentrate on the access to plutonium as being the scapegoat for nuclear proliferation so long as this and other explosive-grade nuclear materials and the corresponding technologies can be sufficiently safeguarded.

CONCLUSIONS

It must therefore be one of INFCE's most important objectives to promote nonproliferation, with full use being made of a closed U-Pu fuel cycle. Whatever technical improvements are available should be used, so long as the cost-effect relationship remains in reason, which means that the technical soundness and cost of technical fixes like coprocessing, coprecipitation, and other suggestions have to be judged under nonproliferation measures like expectable increase of warning time by the introduction of such fixes. In the view of quite a number of experts, there is no doubt that the amenability to surveillance of all types of nuclear fuel cycles is not a question of fundamentals but one of technology: The purpose of the safeguards—namely, timely discovery of diversion of significant quantities of nuclear explosive-grade material or of other materials assisting the production of the same—can be effected for all materials in question and for the associated technical facilities—that is enrichment, reprocessing, plutonium processing, and heavy water production plants. The various measures used will cover comprehensive and audited bookkeeping, the employment of interlock techniques, the protection of material by containment methods with highly sensitive measuring devices to detect any physical barrier bypassing, and continuous inspection, depending on the type of materials and systems. It would be wrong to criticize the safeguards system by alleging that it can only discover and not prevent abuses, as it cannot be assigned the latter function. Due to the large quantities of plutonium already existing in one form or another (including nuclear weapons) in the world of today, effective safeguards and physical protection measures are required anyhow, independent of whether one defers or promotes the plutonium economy.

International solutions can provide an additional nonproliferation element in contrast even to those national solutions that are above suspicion. However, the road to such solutions can only be taken step by step and on the basis of trust—with one side taking the advance of trust if need be—which the stronger countries must show to the weaker in the propagation and assurance of nonproliferation objectives. The problem of multinational solutions is not so much a question of fundamentals but rather a question of the time, patience, and growing experience with which arrangements must be negotiated between a number of partners on legal status, administrative and management structure, financing criteria, and commercial conditions for the services rendered by such enterprises, along with a host of other details. Such solutions cannot be thought out at the conference table but must develop out of practical experience. Furthermore, it is decisive to take into account that multigroup solutions will be stable only if they serve the interests of all parties involved. From the nonproliferation point of view it is more important that national solutions which exist already or will be established in the years to come have from the very beginning an international character as far as their safeguards system is concerned. This is the case where IAEA safeguards will be applied.

It is no doubt to the U.S.'s credit that they once again have drawn the world's attention, on the brink of nuclear energy's breakthrough as the world's major power supplier and to the problem of nuclear proliferation, to possible weaknesses in the safeguards and surveillance system to be applied and thereby have spurred the search for suitable solutions. The approach of endeavoring to find common solutions to those problems that affect all countries is certainly of great importance. But this makes it all the more difficult to understand why the attempt to gain international consensus has been upset by unilateral stipulation beforehand by confronting the world with the U.S. Nuclear Non-Proliferation Act 1978 in its existing form. In setting up INFCE, it was understood by all sides that during the two years or so duration of this study, no new measures were to be undertaken that would jeopardize programs already under way or existing agreements on the peaceful use of nuclear energy.

The danger of the U.S. policy on nuclear energy and nonproliferation is that its overemphasis—although no doubt made with the best of intentions—of the proliferation risk presumed to attend nuclear energy will cause solutions to this highly political problem to be mainly sought in the field of peaceful use of nuclear energy. Another danger is that by overemphasizing the need to introduce a new nonproliferation regime, the existing regime will be devaluated and that the emphasis on the need for a new regime and alternative reactor systems will be used in justifying the objective to gain time by deferring the onset of the plutonium economy and by refusing to make available "sensitive" fuel cycle equipment or technology in order to keep nuclear energy users dependent on the owners of these technologies. To compensate this dependency by supply guarantees is a doubtful solution. Apart from the fact that these guarantees were to comprise the front and the back end of the fuel cycle as well, the trust by recipient countries in such guarantees has been lost in the past, and it is difficult to see how this confidence can be restored again, unless by confident cooperation under nondiscriminatory conditions.

Doubts also arise with respect to the setting up of an international nuclear fuel authority. Were this body to have a worldwide supply monopoly, this would be the end for free market activities in the fuel cycle field. The services rendered by an international nuclear fuel authority under acceptable, nondiscriminatory conditions could well be useful for newcomers or in case of emergencies; but as a means of keeping customer countries dependent via supply conditions that involve renunciation of self-sufficiency or special veto rights, such authority would cause more problems than it would solve.

Nonproliferation remains primarily a political task that must be fulfilled by political means. These include active diplomacy in the cause of peace, with progress being made in disarmament negotiations, safety guarantees underwritten by the major powers, the maintaining of alliances and political stability, and disincentive to possess nuclear weapons. It should be made clear to all nations that any chance of gaining an "advantage" by transgressing against the exclusively peaceful use of nuclear energy would in the last resort in any case be less than the certain damage. On the basis of regained trust and confidence, there is every prospect of success in endeavoring to have as many countries as possible become nonproliferation treaty signatories or at least have the sense to accept of their own free will equivalent safeguards for all nuclear materials and facilities on their territory (full scope safeguards) without the fear that this would weaken their position vis-à-vis other states. Experience shows that well-informed decisions by sovereign states bring about better and longer lasting solutions than temporary submission by an importing country against its better judgment under the pressure of short-term needs. Not without reason has INFCE been emphasizing the need for a balance between energy supply and nonproliferation, in the knowledge that the risk of an energy crisis is as great a threat, if not greater, for the safety of mankind than proliferation. It should be borne in mind that to prevent proliferation, pragmatism and flexibility can be more useful than rigid universal principles if certain basic guidelines are met.

In this context, it can only be wished that the major uranium-exporting countries realize, when fixing their export legislation with a view to nonproliferation, that these conditions must remain within a balanced framework, acceptable to the importing countries as well, if they are to be effective. Misuse of an exporter's position to push through unjustified political objectives does not—as the oil crisis showed—lead to stable solutions but provokes the risk of undesired counterstrategies.

It would greatly hamper cooperation between the supplier countries if there were to be undue divergence in their respective views on the correct approach toward the common goal of exploitation of nuclear energy while preventing proliferation. It would be fatal for the future development of nuclear energy and its peaceful use for the benefit of all if the raw materials, technologies, and services it requires were to be made available only under the restrictive conditions of monopolies or cartels, partly or not at all.

Chapter 5

Nuclear Policy: The U.S. Approach to Nonproliferation

*Joseph S. Nye, Jr.**

I have been asked essentially to give a "midterm grade" to the Carter administration's efforts to slow the spread of nuclear weapons. The task is difficult for two reasons. First, as a government official, I have access to information not available to the public; at the same time my role as a government participant is a possible source of bias. Second, we are trying to make a short-term assessment of what is by definition a long-term process. Moreover, progress has to be judged in the light of estimates of what otherwise would have been the situation.

Obviously there is no neat solution to these difficulties, but a good way to start is by making clear what the U.S. government is trying to achieve. The goals of our nonproliferation policy are to slow the rate of spread of nuclear weapons, preferably to zero, and to construct a stable international regime for the governance of nuclear energy. These goals can be judged by whether the administration efforts have contributed to a rate of proliferation lower than it otherwise would have been, have promoted a nuclear fuel cycle that is more proliferation resistant than it otherwise would have been, and have strengthened institutions for a stable international regime. Later I will provide evidence of significant progress on each of these dimensions, but first I wish to clear away

* Deputy to the U.S. Undersecretary for Security, Assistance, Science, and Technology

some misunderstandings of our policy and describe what I think would have happened in the absence of the new U.S. approach.

Some critics have charged the Carter administration with failing to see that proliferation is a political problem and seeking a technical fix through abolition of reprocessing. They argue that the peaceful nuclear fuel cycle is not a source of proliferation because there are more efficient ways to develop a weapon. Thus, in their view, the American initiatives have simply created turmoil, reduced American exports, isolated the United States, and created incentives for proliferation.

It is true that the more newsworthy Carter initiatives have focused on the fuel cycle, but it is not true that the political dimensions have been ignored. We have always regarded proliferation as basically a political problem. What is more, we have not regarded the fuel cycle as the largest part of the problem. But neither is it a trivial part. For example, a recent General Accounting Office report issued strong support for the Carter administration's view that large commercial reprocessing plants with inadequate safeguards present a greater proliferation risk than small clandestine plants. In addition, let us hope that recent press revelations will finally lay to rest the spurious argument that because there are more efficient ways to develop weapons than through misuse of the fuel cycle, no state would misuse the fuel cycle. That a priori argument, heard so frequently in the past, has proven to be the real example of a technical case without political context.

MEASURES TO DEAL WITH INCENTIVES

The proliferation problem has both a supply and a demand aspect. Sound policy has to address both the supply of capabilities and the demand for weaponry. The fact that policy measures focused on capabilities have attracted more recent publicity does not mean that policy measures addressed to incentives have not been given equal weight internally. As Sherlock Holmes once noted, the fact that a dog does not bark in the night may be the more important clue. In practice, we regard the security guarantees that the United States provides to its allies as our most important nonproliferation policy instruments. Critics miss this point when they complain that the administration failed to pursue disputes over reprocessing with our allies because it feared to destabilize the alliances. Any policy pursued to the point of severely shaking those alliances would be a failure in nonproliferation terms. A cooperative approach with our allies is not only good alliance policy, it is also good nonproliferation policy.

Similarly, we have been concerned to protect the multilateral instruments that have been laboriously constructed over the past two decades to address security motivations. Most important, of course, is the Nonproliferation Treaty (NPT), which 105 nations have now ratified. The treaty has helped to create an international regime in which states agree that their security interests can be better served by avoiding the further spread of the bomb. It provides important reassurances that potential adversaries are confining their nuclear activities to peaceful purposes.

The NPT is a delicate international arrangement. Countries without nuclear weapons have accepted an explicitly unequal status in the military area on the condition that they be treated equally with regard to civil nuclear cooperation. Thus we have rejected a number of suggestions for policies on the civil side that would have weakened the fabric of the treaty as one of the key nonproliferation institutions.

Another multilateral instrument is the nuclear-weapons-free zone. The most important example is the Latin American nuclear-weapons-free zone, which was established in the 1960s by the treaty of Tlatelolco, but which lacked several adherents, including U.S. ratification of its first protocol, before becoming fully effective. Early in his term and without much fanfare, President Carter announced that the United States would ratify the protocol. Subsequently Argentina declared its intent to ratify the treaty, and the USSR and France announced intentions to ratify the relevant protocols. Only Cuban action will be necessary before the treaty enters fully into force, and even that precondition could be waived. Finally, American efforts to control the vertical proliferation of nuclear arsenals through the Strategic Arms Limitation Talks and comprehensive test ban negotiations have an important indirect effect on nonproliferation incentives. In short, there has been significant, if less noticeable, progress relating to incentives.

EFFORTS TO SEPARATE PEACEFUL FROM MILITARY CAPABILITIES

Incentives can be reduced, but they cannot be eliminated as long as national rivalries and security concerns exist. We must also deal with capabilities to develop nuclear explosives. The fact that civil nuclear technology and material can be used to develop nuclear weaponry has presented a dilemma that we have recognized since 1945. We have gone through four phases in our efforts to limit the spread of nuclear explosive capability. The first was the Baruch plan to create a strong international authority to develop nuclear energy. It was a more ambitious step than international realities at the time would permit. American policy then turned to a posture of seeking to protect its monopoly by severely restricting the export of any nuclear technology. In December 1953, President Eisenhower launched a third approach with his Atoms for Peace program. The idea of the Atoms for Peace approach was to assist countries in their development of civilian nuclear energy, in return for their guarantees that they would use such assistance only for peaceful purposes and under safeguards.

In practice, the early Atoms for Peace policy failed to achieve the right balance, but its philosophy made sense as a long-term strategy. Essentially, the United States was offering to share the fruits of its then long technological lead at an accelerated pace in return for the acceptance by other countries of conditions and institutions designed to control any destabilizing effects from such sharing. Specifically, the major accomplishments were the institution of a system of international safeguards administered by the International Atomic Energy Agency (IAEA) and later the Nonproliferation Treaty, which came into force in 1970.

In the early 1970s the proliferation situation seemed quiescent, but complacency was shattered by two events that ushered in the fourth period of turmoil that has been with us since 1974. One was the Indian explosion of a "peaceful" nuclear device using plutonium derived from a Canadian-supplied research reactor—an event viewed as violating the spirit if not the letter of the loosely written 1950s vintage Canadian-Indian agreement. The Indian explosion gave rise to strong demands for stricter export policies in both the Canadian Parliament and the U.S. Congress.

The other big event was the oil embargo and fourfold increase in oil prices, which created widespread insecurity in energy supply. Problems with oil led to a resurgence of expectations about the importance of nuclear energy and raised questions about the sufficiency of uranium fuel. This was exacerbated by the 1974 decision of the Atomic Energy Commission to close the order books for enrichment until they could be certain that supply would equal demand. The net effect was to stimulate independent enrichment projects—incidentally, long before President Carter came into office.

Another effect was to reinforce plans for early commercial use of plutonium fuel. In several troubling cases, reprocessing plants were ordered by countries before they had built their first thermal reactor. Moreover, the IAEA projected that some forty-six countries would have reprocessing needs by 1990. All this would occur before appropriate technology and institutions had been developed. The implications for the fragile regime of international safeguards threatened to be disastrous.

Recovering from a late start, the Ford administration undertook important initiatives in 1975–1976. It began to organize the nuclear supplier governments to agree on a code of conduct for nuclear exports. And in the final days before the 1976 election, President Ford announced a moratorium on commercial reprocessing of spent fuel in the United States pending further evaluation. At the same time, a number of congressional initiatives were undertaken to tighten the conditions for nuclear exports from the United States, and several private studies of the nuclear fuel cycle—notably the so-called Ford-Mitre report and the American Physical Society report—were coming to the conclusion that the commercial use of plutonium was economically premature and potentially dangerous.

When the Carter administration came into office, there was a widespread but by no means universal perception that the spread of sensitive nuclear facilities (particularly uranium enrichment and reprocessing) and the planned early and wide-scale use of plutonium as a nuclear fuel threatened to erode the delicate instrument of the IAEA safeguards system and to make increasingly porous the barrier between peaceful and nonpeaceful applications of nuclear energy. The new administration did not create the period of turmoil in international nuclear cooperation. Rather, it inherited a highly unstable situation. Another major setback to nonproliferation might very well have brought the end of the international regime so laboriously constructed in the 1950s and 1960s. The task before us was to restore and strengthen a regime that would balance legitimate energy requirements and nonproliferation concerns.

The administration recognized that there was no single technological fix that would create a safe fuel cycle but rather sought to move toward a series of technological and institutional steps that would lessen the risks while allowing legitimate energy needs to be met. To gain the time necessary to develop technological and institutional arrangements, the administration urged that premature commercialization of fuel cycles utilizing plutonium be avoided and announced that the United States, for its part, would defer its own plans for commercial reprocessing and recycle of plutonium.

The administration was and remains strongly against recycle of plutonium in thermal reactors as posing a clear and present proliferation danger in return for, at best, marginal economic and supply assurance gains. Breeder reactors, however, are a significant potential long-term energy alternative, and we have been careful not to oppose breeder research and development programs at home or abroad. We have, however, expressed reservations about their commercial deployment before proliferation-resistant technological and institutional alternatives are investigated.

We recognized that we could not unilaterally impose our will on others concerning how the nuclear fuel cycle should be structured and that we did not have all the answers ourselves. For this reason, six of the seven points in President Carter's April 7, 1977, nonproliferation statement dealt with issues within our domestic jurisdiction. The seventh point was to lay the basis for the development of an international regime of norms and institutions that will provide the widest possible separation between peaceful applications and potential military uses while enabling countries to meet their energy needs. A key element in bringing about such a development was the suggestion for an International Nuclear Fuel Cycle Evaluation (INFCE). The idea of INFCE evolved from the prior administration's reprocessing evaluation program. The Carter administration broadened this idea to include other nations and to encompass all aspects of the fuel cycle, not just reprocessing.

INFCE has been described as a pioneering effort at international assessment. Certainly, the United States sees INFCE as a cooperative effort to evaluate the role of nuclear energy technology and institutions in an international context and help develop an objective appreciation of the nonproliferation, economic, and other implications of different fuel cycle approaches. INFCE provides a two-year period in which nations can reexamine assumptions and search for ways to reconcile their somewhat different assessments of the risks involved in and the time scale for commercialization of the various aspects of the nuclear fuel cycle. While INFCE has a predominantly technical cast, it is part of the political process of laying a basis for a stable international regime to govern nuclear energy through the end of the century.

While it is too early to predict the outcome of this two-year assessment, the United States has indicated, in broad outline, the type of political solution that we believe can bring an end to the period of turmoil over the nuclear fuel cycle issue. A stable regime should be designed to minimize the global distribution of weapons-usable materials and to reduce the vulnerability of sensitive points in the fuel cycle, while adequately meeting the energy needs of all countries. As I suggested in my speech to the Uranium Institute in London earlier this year, we envisage five basic norms for a strengthened international regime:

1. Full-scope safeguards,
2. Avoidance of the unnecessary spread of sensitive facilities,
3. Use of diversion-resistant technologies,
4. Institutionalized control of sensitive facilities, and
5. Institutions to insure the availability of the benefits of nuclear energy.

DECREASING THE RATE OF PROLIFERATION

I would now like to assess the progress that we have made in light of the three tests I mentioned earlier. The basic test is whether the Carter initiatives have caused the rate of proliferation to be higher or lower than it otherwise would have been. While I cannot get into the specific cases that support my conclusion that it will be lower, I believe a good case can also be made in general terms. Basically, proliferation is less likely to become a cheap option. If nothing else, the high priority that the Carter administration has given to the issue has raised second thoughts among those who might have wanted to approach the option because it cost little.

The attention of both suppliers and consumers has been called to the dangers of proliferation. The very vocabulary at INFCE meetings—"proliferation resistance" and "weapons-usable materials"—indicates change. The publication of the Nuclear Suppliers Guidelines earlier this year, including the provisions for safeguards, for special restraint on sensitive exports, and for supplier consultations on possible sanctions if recipient countries violate safeguards, have made it apparent to a potential proliferator that questionable activities are unlikely to go unnoticed and that there are likely to be significant costs involved in "crossing over the line."

Section 307 of the U.S. Nonproliferation Act of 1978 reinforces this by requiring termination of U.S. nuclear cooperation to states that detonate a nuclear explosive device; abrogate, terminate, or violate safeguards; or engage in activities directly related to manufacture or acquisition of nuclear explosive devices. The deterrent effects of international safeguards are a function of the likelihood of detection together with the cost of ensuing sanctions. In effect, the sanctions aspect—and thus the deterrent effect—of safeguards has been strengthened over the last two years.

PROMOTING A PROLIFERATION-RESISTANT FUEL CYCLE

The second test is whether the fuel cycle will be made safer than would otherwise have been the case as a result of the U.S. initiatives. In terms of what has been accomplished in developing a consensus on a more proliferation-resistant fuel cycle, the U.S. approach has stimulated a general reanalysis of long-held assumptions and a reconsideration of previously rejected alternatives. A number of key governments are now studying options to increase proliferation resistance rather than proceeding on a "business as usual" basis. Industry at home and abroad has also begun to look at ways to reduce proliferation risks.

In more specific terms, we are beginning to see a reconsideration by a number of states of the need for recycle of plutonium in thermal reactors. If this develops into a near consensus, it will mean that plutonium in large quantities will not be needed until the breeder is ready for commercial deployment, which, for the vast majority of countries, is decades away. This provides additional time to reduce the risks associated with these reactors and/or to develop alternatives and strengthened institutions.

Reprocessing and the breeder are not, of course, the only vulnerable points in the fuel cycle. We must find technical and institutional combinations to reduce the dangers at each potentially sensitive point in the fuel cycle. At each point, there are technical and institutional choices that present different degrees of resistance against diversion and seizure of weapons-usable materials from peaceful nuclear activities.

There are three basic components to reducing vulnerability—economic justification, technical design to minimize risks and assure effective safeguards, and international institutional arrangements such as joint or multilateral control. The appropriate "mix" of these components will depend on the kind of activity involved and the circumstances of the specific case involved.

Activities associated with short times from diversion to weapons development and/or with low detectability will need additional international frameworks to be considered safe. If there is a sound economic justification—but the activity and the technical design do not assure effectiveness of safeguards—institutional arrangements such as multinational control should be a prerequisite for going ahead with the activity. Of course the matrix covering all the possible combinations is complex, but a series of prudent choices at each sensitive point in the fuel cycle can contribute significantly to the goal of maintaining the distance between peaceful and military uses of nuclear energy that otherwise very likely would have eroded.

In INFCE and elsewhere, considerable progress has been made in some of the above areas, such as reducing the risk associated with the use of highly enriched uranium in research reactors. Solutions to others will be more controversial because they are likely to involve added safeguards, costly technical modifications, or the creation of institutional arrangements. Discussions now being carried on in INFCE relate to all of these points and appropriate "mixes" for each activity. We have noticed with interest efforts by other governments to suggest the broad outlines of solutions. For example, the delegate of the Federal Republic of Germany recently told the IAEA general conference:

> There is a growing feeling that specific technical amendments of isolated institutional arrangements will not solve the existing problems. It rather appears to be desirable, and also possible, to identify a bouquet of coordinated measures which at the end of the evaluation might be submitted with a high degree of consensus to the Governments for their decisions. Without prejudging the further development, one might expect to find among these measures some of the following items:
>
> further technical development safeguards;
> increasing reliability of fuel supply for nuclear power stations;
> criteria for the use of highly enriched uranium in research reactors and new reactor types;
> closer investigation of possible modifications in some current back-end of fuel cycle technologies;
> establishment of a regime for the deposit of excess plutonium as provided in the Agency's Statute;
> mechanisms for international or regional institutional cooperation.

STRENGTHENING INSTITUTIONS

The third measure of progress is how we are doing in strengthening existing norms and institutions and developing new ones for building up the international regime. International safeguards administered by the IAEA are of course the fundamental norm, and progress has been made both in strengthening the effectiveness of these safeguards and in expanding their application. For the first time the agency issued a safeguards implementation report that addressed problems that it has encountered in carrying out its responsibilities. This report will be undertaken annually, and work is already under way in remedying the deficiencies that have been identified and in developing methods to safeguard new types of nuclear activities.

The avoidance of commercial competition that would weaken the application of safeguards has been assured by the Nuclear Suppliers Guidelines. Moreover, several countries, including the United States, have adopted a requirement that a recipient country have all its nuclear activities under international safeguards as a condition of nuclear supply (full-scope safeguards). Only a handful of countries do not meet this standard, and after informal consultation with other governments, we believe that there is a good prospect for widespread acceptance of such safeguards by both suppliers and recipients at the end of the INFCE period.

Beyond strengthening the present safeguards regime, we have begun to develop institutions to implement the principle of assurance of benefits. Supply assurances (such as a fuel bank) and international spent fuel repositories are examples of institutional arrangements that can reduce the incentives for countries with small programs to develop unnecessary enrichment and reprocessing facilities. We have been pleased by the initial positive responses to the idea of a fuel bank consisting of a stockpile of fuel to be released to countries that have all their facilities under safeguards, have a clean proliferation record, and have chosen not to develop sensitive facilities on a national basis.

Development of international spent fuel storage regimes is also important. For some states, long-term away from reactor storage at home is not a viable alternative because of political, environmental, and geological considerations. The United States has indicated its willingness to take a limited amount of foreign spent fuel for storage in the United States and is engaged in the discussion of international repositories in working group 6 of INFCE.

In addition to these institutional arrangements designed to reduce the incentives and concerns that would lead to premature development of sensitive nuclear facilities, we have begun studies and discussion of institutions for effective joint control of these sensitive facilities that are economically essential and difficult to safeguard nationally. This is particularly applicable to enrichment and reprocessing and perhaps plutonium storage regimes where some type of multinational ownership and management and possibly new rules of operation might help reinforce the effectiveness of international safeguards. Discussion of such possible arrangements is under way in INFCE, and we will devote increasing attention to this as we work toward a consensus on managing the fuel cycle.

CONCLUSION

In short, I believe that we have seen credible progress on each of the three crucial measures discussed above. On the other hand, there are those who argue that whatever short-term progress we have achieved in controlling proliferation, the "restrictive" policies we are pursuing will over the longer term trigger a rush to independent nuclear fuel cycle capability, undermining the interdependence of the international fuel cycle and thereby reducing the barriers to proliferation.

This line of argument represents a rather fundamental misunderstanding of U.S. policy. We do not seek to "turn off" the development of sensitive technology necessary to meet present and projected energy requirements or to delay the deployment of facilities embracing such technology when there is a clear economic justification for them. We do not have the leverage to accomplish this even if it were our objective. We do believe, however, that the number of sensitive facilities should be limited to those necessary to meet actual energy requirements and that appropriate safeguards and institutional arrangements are legitimate "costs" that must be factored into the development of these facilities. Our approach is evolutionary rather than prohibitory.

Avoiding premature spread of sensitive facilities that involve weapons-usable materials is a common interest of nations that want a stable international regime. To support premature spread before safer technology and institutions have developed works against the greater interest. It is worth remembering the criterion set for nineteenth century hospitals: "At least they should not spread disease!"

Moreover, I see little evidence that the new U.S. approach has stimulated development of additional national facilities. Reprocessing plans were underway in the United Kingdom, France, Japan, the Federal Republic of Germany, and other countries well before our policy was formulated, as were arrangements to transfer this technology to other states. On the contrary, I note that these four countries have indicated a willingness to consider technological and institutional modifications in the interest of nonproliferation and that France, Germany, and the United Kingdom have all announced that henceforth they do not expect to export reprocessing plants. As I indicated above, we are beginning to see similarity as well as differences in the many discussions in INFCE and elsewhere on ways to meet energy needs without increasing the risk of proliferation.

To summarize my evaluation of the progress over the last two years: First, the recent U.S. initiatives have increased recognition within the international community of the costs involved in "crossing the line" from peaceful nuclear activity to nonpeaceful applications. Proliferation is less likely to become a cheap option. This has added to the deterrent effect of the international safeguards system.

Second, there has been a heightened awareness of the dangers of continuing development of the nuclear fuel cycle based upon past assumptions, and an increased readiness to reexamine these assumptions and to look for alternatives.

Third, international and domestic evaluations of these alternatives have been undertaken and continue. We recognize that there is no single answer, either technological or institutional, to the problems we face. We cannot look for a completely "risk-free" nuclear fuel cycle. However, we can reasonably expect a series of improvements in various aspects of the fuel cycle that will add up to a significant gain in preventing erosion of the barriers between peaceful application of nuclear energy and nonpeaceful uses.

Fourth, rather than sitting back and accepting erosion in the face of technological change and spread, steps have been taken to strengthen the IAEA safeguards system that is central to any nonproliferation regime, and work has begun on other institutional arrangements to complement the safeguards system.

Obviously, these are interim judgments. Controlling the risk of proliferation is and will continue to be a dynamic exercise as we adjust to changing energy requirements, security, and political perceptions and technological developments. The struggle will not be finished by the end of INFCE, not during the life of this administration, and perhaps not in our lifetime. The important thing is that the international community is making a renewed attack on this fundamental issue. I believe that there is considerable hope that we will find ways to insure that this essential technology continues to serve mankind rather than to threaten it.

Chapter 6

European Views on Nonproliferation

*René Foch**

THE PERIOD OF EUPHORIA

The Atoms for Peace program, one of the earliest steps in the direction of a controlled nuclear development, was consonant with the American philosophy in general and with the way many Europeans view the United States. There was a parallel with the Marshall Plan philosophy. It was, I think, not only a generous policy but—what is even more important when talking about politics—it was an intelligent policy. In other words, in recognizing that there are things you cannot prevent, it was a policy of helping development in such a manner that this development takes place in the most responsible manner. There was a period in eighteenth century European history when people talked about enlightened despotism. In retrospect I would qualify the Atoms for Peace program as a form of "enlightened monopoly." It gave birth, in particular, to a unique form of American-European cooperation, and although the signature of the Non-Proliferation Treaty in 1968 made for tough negotiations to reconcile the original aspect of U.S. policy (I am referring to Europe in particular) and the worldwide aspect, on the whole the Non-Proliferation

*Former Director, Euratom

Treaty achieved very important results and a kind of international consensus.

I would now like to draw your attention to a clause of the NPT that was an essential condition for its acceptance by certain countries. I am referring to the famous Article 4. For instance, the signature of the URENCO agreement by which Germany, England, and the Netherlands agreed to form a firm to build plants to produce U-235 took place on the same day, I think, of the coming into force of the Non-Proliferation Treaty, and indeed, at that time the U.S. position was unequivocal. Arthur Goldberg, at that time American Ambassador to the United Nations, said exactly that in May 1968:

> The point is that there is no basis for any concern that this Treaty would cause inhibition or restriction of the opportunity for a non-nuclear-weapons state to deploy its capabilities in nuclear science and technology. The whole field of nuclear science associated with electric power production is accessible now and will be more accessible under the Treaty for all who seek to exploit it. This includes not only the present generation of nuclear power reactors but also that advanced technology which is still developing fast breeder power reactors.

Indeed I remember taking part in the negotiation of an exchange of information between the European programme on fast breeder and the U.S. programme, and when that contract was finally solemnly signed we got a telegram of felicitation from the then Secretary of State, who congratulated us on this cooperation in the forefront of nuclear technology, thus providing a concrete example of the Atlantic partnership in action.

THE PERIOD OF DISAGREEMENT

Then a number of disagreements started, particularly after the oil crisis made it clear that nuclear energy, which had so often been said to be just round the corner, was now serious business. Indeed, it is sometimes forgotten, but one of the first U.S. reactions to the oil crisis was the launching of the so-called Project Independence, which included an important nuclear element. Thus, the question of the supply of uranium—and, of course, of U-235—became just as vital as the question of oil supply, and the United States inadvertently underlined the importance of this point when, in 1974, it suddenly turned away orders for future enrichment of uranium, to which EURATOM, Brazil, and some other countries had committed themselves and for which, incidentally, they had partially paid in advance. Thus the lesson was learned that outside dependence, even on friends' U-235 supplies, presented serious problems for long-term energy planning.

The next year, 1975, things were made worse by a temporary administrative embargo on reactors and fissile materials by the U.S. administration, and this time the European Atomic Energy Community felt that the matter was so important that it made a formal protest to the U.S. authorities. This decision, taken without any consultation, and which therefore the European commission is in no position to explain to the various users in the community, is threatening the normal development of the nuclear programs of the community and throws grave doubts on the viability of supplies coming from the United States. This was the first sign of a worsening of relations between the United States and Europe. The situation is all the more regrettable since the community has just adopted a program of development of nuclear energy that aims to reduce, as far as possible, the oil imports of the community—an aim supported by the U.S. government and also by the other oil-importing industrialized countries.

Then there was the case of India. Regardless of whether India did break or did not break an international agreement in making an explosive device, it does not make sense that as a result the European countries have been deprived of their supply of U-235 or, in the case of Canada, of natural uranium. That reminds me of an episode: out of one hundred military cadets on leave, ninety-eight came back on time and two did not. The ninety-eight who returned on time were punished. That is about the same kind of logic.

The third event that played an important part in the worsening of U.S.–European relations in the nuclear field was the German–Brazilian deal. In this situation, the various branches of the U.S. government reacted in different manners. Starting in the fall of 1974, at the suggestion of the U.S. government, Britain, France, Japan, Germany, and the Soviet Union, and later Belgium, the Netherlands, Italy, Sweden, Switzerland, Czechoslovakia, East Germany, and Poland, met several times in the so-called Group of London. Finally, by November 1975 they agreed on certain guidelines that were made public in January 1978. These guidelines stipulate that governmental assurance should be given explicitly, excluding uses that would result in any nuclear explosive, that nuclear material was to be adequately protected against theft and strictly safeguarded. The U.S. government failed at that time to get the so-called full-scope safeguard system for all facilities in recipient countries. Politically most important, however, suppliers promised restraint in the transfer of sensitive facilities, technology, and weapons-usable material and products to avoid the production of any unsafeguarded nuclear material. These policies were explicitly meant not to apply in retroactive fashion to the French-Pakistan deal and to the German-Brazilian deal. Nor did the suppliers accept a clause that would have made the future of processing dependent on prior approval by the supplier group. So here we have an example, and I think on the whole a fairly successful example, of a method to deal with the problem—namely, consultation between executive branches of the various governments concerned.

A second compromise that was inherent to the adoption of these guidelines was the idea of multinational facilities as an alternative to national enrichment and reprocessing efforts. The United States and others had resurrected this approach and had initiated a major IAEA study on regional fuel cycle at the 1975 Non-Proliferation Treaty Conference. The only trouble was that by the time the studies were published, the United States had shifted to a more skeptical position, considering such centers to be disseminators of knowledge and sensitive technology.

The Commission of the European Energy Community was asked to give its opinion according to the regulation of the EURATOM treaty, which opinion was: first, it is favorable to the whole idea. Second, it insists that there is such a thing as the European community and that there is such a thing as a nuclear common market. This means that fissile materials, in particular equipment, can move freely within the community, just as do cars or refrigerators, and that the guidelines adopted in London should never be interpreted in such a way as to jeopardize the free flow of material and equipment within the European community.

Subsequently, a second branch of the U.S. government came into the picture—namely the U.S. Congress—and finally the U.S. Congress adopted a bill that was formally signed by the president of the United States in the spring of 1978 that formulates criteria for new export agreements, significantly tightening previous rules for full-scope safeguards. It includes a ban on explosives and provides for a cut-off of all aid in the event of any retransfer or reprocessing of American-supplied nuclear material without authorization. The bill permits presidential exceptions, which can be overruled by Congress but which require the government to renegotiate all agreements with other countries to conform to the new criteria.

A German study[1] of this legislation and its likely effects on European–U.S. relations in this field makes five points: First, the legislation forces the government to reopen international agreements that were negotiated and ratified earlier. Some countries are unlikely to accept the abrogation of commitments simply on the grounds that the U.S. Congress and administration have changed their minds. Second, the legislation not only regulates bilateral relations between the U.S. and its partners but also between third countries.

Third, the bill forces the United States to exploit its position as a supplier of nuclear technology and uranium in order to impose American ideas about reprocessing on other countries. Many recipients of American uranium will therefore have to draw a blank check to guard against the prospects that the White House and Congress in ten or twenty years will share their attitude on energy or environmental policy that caused them to move toward reprocessing.

Fourth, the legislation attempts to deal with the dilemma of formulating nondiscriminatory, universally applicable nonproliferation rules, while being sufficiently flexible. In the view of most of America's allies, the bill does not succeed in making sufficient allowance for countries like the EURATOM countries or Japan.

Fifth, the special clause concerning reprocessing, as far as Europe is concerned, would apply exclusively to the future German reprocessing plants. Thus, a country that has already undertaken additional commitments of nonproliferation, notably the 1954 renunciation of the production of nuclear weapons, is being discriminated against. Finally, concludes Kaiser, the bill is heavily oriented toward a technological approach to nonproliferation, and we are increasingly aware of how fragile such routes are likely to be.

Kaiser's view is, I think, one of the most thorough and balanced. As for my own view of this legislation, the U.S. government finds itself committed to a global nonproliferation policy through legislative means, and looking at it from a political angle, there are of course historical reasons for this. Traditionally, even in the days of the so-called imperial presidency, Congress, through the Joint Congressional Committee on Atomic Energy, had played a very important role in the nuclear field. Now, of course, with the backlash of Vietnam and Watergate, Congress is becoming even more powerful, but has disbanded this unique committee in which there was quite a lot of competence.

When you are trying to settle a problem such as nonproliferation by law, you must attempt to treat all parties in a nondiscriminatory fashion. I would like to reflect for a moment on this concept of nondiscrimination. To begin with, the very concept of nonproliferation is in itself discriminatory: on the one hand there are the nuclear powers; on the other there are the nonnuclear. So the idea of a plain discrimination by nondiscriminatory means is to my mind—perhaps I am too Gallic about it—almost impossible to conceive. Now, if we are talking about discrimination, there are two ways to discriminate. You can discriminate by treating differently people who are more or less in the same position, as for instance the way the French authorities felt when the British authorities were receiving assistance for their military program and the French authorities were denied similar assistance. But you can also discriminate by treating identically people who are in different positions—for instance, by treating Germany like India. To my mind, if I may make a comparison with the field of foreign trade, it is a kind of less favored nation clause, which is contrary to the very concept of an alliance. After all, the concept of an alliance is precisely to discriminate, to treat your friends better than the people outside or, at least, better than the people against whom the alliance has been conceived in the first place.

There are also geographical and geopolitical reasons, to my mind, why this approach cannot work. In fact, the U.S. Congress is attempting to legislate on other people's territory, using what is left of the U.S. monopoly as leverage. Now this is flatly contradictory to two basic principles of American history, as I read it—a political one and an economic one. The political principle was joined very clearly at the famous episode called the Boston Teaparty. The House of Commons had voted certain taxes on tea that the people of Boston did not like, and they threw the stuff into Boston Harbor. A second principle of U.S. economic philosophy is that monopolies are bad, so I think that this piece of U.S. legislation is so foreign to the whole trend of American society that it cannot work. There are also judicial reasons, which were alluded to by Karl Kaiser—for example, the conflict with previously signed international agreements. Finally, there are technological reasons in a field where facts change so quickly. If you are trying to regulate the field by such cumbersome procedures as law, you are always running behind the scientists.

Thus the conclusion I would draw from these reflections is that it is far better to have consultations between executive branches along the lines of the London Group and, when the need arises, to have executive agreements with revision clauses. Here I would like to draw your attention to a legal instrument that was negotiated in another field of advanced technology—the Intelsat agreement. When the Intelsat agreement was first negotiated, the U.S. position was technically absolutely overwhelming, and the first draft presented by the U.S. lawyers would have, so to speak, frozen forever this position of imbalance. Europe then said, We need those satellites, so we go along with the idea and we agree that CONSAT will technically run the show, but we do not agree to this state of imbalance, which is likely to be corrected soon, so we will sign a provisional agreement with a revision clause in five years. Perhaps we can draw a leaf from that particular book.

THE EUROPEAN VIEW

Concerning the important question of reprocessing and breeder development, the European Atomic Energy Community has taken an official stand in stating that the community and its member states must keep open the possibility for building fast breeders, starting in the year 1990, and that breeding must be pursued with the utmost vigor, since the figures show the importance of breeding for the Europeans. The clash between the two approaches was made very clear, although it was not a public event, at a meeting of the so-called Trilateral Commission in Bonn in October 1977. The European viewpoint was presented at a special debate on nonproliferation.

The European viewpoint, presented by M. Giraud, at that time Administrateur Général of the Commissariat de l'Energie Atomique, and agreed to word for word by the whole European Delegation, reads as follows:

> European countries consider as unacceptable the recent demands of certain uranium-producing countries toying with the idea of forming a political cartel, not because they require commitments on peaceful utilization, which is quite natural, but because they want international rules of non-proliferation to decide, in place of the European Governments concerned, the use that will be made of uranium in the energy balance of their countries. Such blackmail on uranium, if I may say so, would constitute a decisive incentive, if it was needed, towards reprocessing and fast breeder reactors. Similarly, some countries, including France, feel that the system called full fuel cycle safeguards, which uses nuclear proliferation as a means to oblige a country to put under its national safeguards even the activities it has developed by itself, will lead such a country to develop its own programme on a purely national basis, that is, free of any safeguards. This will increase, and not decrease, the risk of nuclear proliferation in that country and the others.

The Japanese viewpoint was more subtle but equally clear: The U.S. is, I should say, unique among industrialized countries of the free world in its possession of energy resources. The U.S. still has an impressive production of oil and gas. Besides the U.S. is extremely well endowed with coal that can easily be obtained by strip mining. Therefore, the U.S. has energy options which are unfortunately not available to most other industrialized nations. Japan is the extreme example of an industrially most advanced country with a relatively large population, that has absolutely no indigenous energy sources. For Japan, the breeder is not

> only a question of economics, it represents the only practical way to decrease its energy dependence to a tolerable level. It is on this long-term requirement of energy resources that the need for reprocessing is based. The energy potential of uranium will be fully realized only if the uranium and plutonium present in spent nuclear fuel is recycled.

So, the clash was looming. At the European Council, the name of the European summit meetings, in spring 1978, the heads of state decided to let pass without making any move the deadline of the U.S. Non-Proliferation Act by which various countries, in particular those of the European Atomic Energy Community, were supposed to communicate to Washington their readiness to renegotiate the U.S.–EURATOM agreement. A compromise was finally reached, on the following basis: Since in the meantime President Carter has suggested the INFCE exercise, clearly it would make no sense to prejudge in bilateral negotiations the results of this multilateral exercise, and therefore the compromise was that—since it is necessary according to the letter of your law—we are ready to renegotiate the U.S.–EURATOM agreement, provided these renegotiations do not touch any of the subjects that are under study in the present INFCE exercise. That is more or less where we now stand.

So far, I have given you the view of European countries as importers, but now there is a growing number of European countries who are able to export and are indeed more and more active in the export market. That means that the Europeans who in the past could be happy with either accepting or refusing U.S. decisions have to take their own responsibilities. From an institutional angle I would say that, *grosso modo,* the attitudes of European countries as importers are covered by the EURATOM treaty, and therefore, the machinery of the European Atomic Energy Community produces more or less agreed European positions.

EUROPEAN APPROACHES TO NONPROLIFERATION

Switching to the field of exports, it is only in the framework of either bilateral conversation—for example, the Franco-German conversation—or in the framework of the so-called political cooperation that an attempt is made to coordinate European views. In most European countries these things are pretty much in the hands of executive branches and I would assume, subject to correction, that most European countries would agree that the best practical way to tackle this problem is through consultation, through the drafting of agreed upon guidelines and the application of these guidelines on a case-by-case basis.

A further idea, which was put forward in particular by Valéry Giscard d'Estaing in his speech on disarmament at the United Nations, is the idea that one must take fully into account the reality of regional situations. He made that point in the framework of general disarmament, but I think it is at least equally valid in the nonproliferation field. For instance, if we apply the idea of giving a premium to regional action rather than looking immediately for a worldwide solution. The idea of a fuel bank, for instance, is to my mind a nonstarter, because either it would in fact be run by the U.S. government—and in that case what is the use—or it would be genuinely international—and I wonder if technically it would work. Seeing, for instance, the amount of administrative and managerial trouble we have had with the European community's uranium supply agency, I have difficulty in visualizing the possibility of running such a fuel bank efficiently at the worldwide level.

The idea of such a fuel bank, and it is made very explicit in the foreword of the U.S. working document, is a candid recognition that the United States did not constitute by itself a reliable supply. Now when this came upon us, there was a reaction, we turned toward the Soviet Union, and you may be surprised to hear that exactly 50 percent of the fissile material used in the European community comes from Russian sources. That is a staggering figure, and it seems that for Europeans the way out of this dilemma would be to constitute in Europe a third center of production, including, when reasonable, certain countries of the Third World that have needs for atomic energy. I refer, for instance, to the participation of Iran in the EURODIF syndicate. Indeed, it would reflect the interlocking of European and Third World country interests, as such an alternative center in what one used to call the free world would provide a kind of fallback position in case for one reason or another the U.S. government could not or would not want to supply fissile materials to a given country. This may take place with the full blessing of the U.S. government. In an example borrowed from another field: an amendment was passed by the U.S. Congress preventing the U.S. government from supplying the Turkish army with weapons in connection with the whole Cyprus business. A solution was found, with the full agreement of the U.S. government, whereby it was possible for the Germans to send second-hand tanks to the Turkish army, with the Germans being resupplied by the United States.

Such multinational enterprises should manage enrichment and reprocessing centers. They might also cover the management of spent fuel. We must be clear, I think, that further on in the future there will be such centers in the Third World as well, in Brazil for instance, so that interdependence, consultation, and responsibility will go together.

In touching upon this question of the Third World, I think it is worthwhile to look for a moment at the Indian case—not at the explosion itself and its political consequences, but at the way India was treated after the coming into force of the nonproliferation act last spring. An article in *The Economist* a few months before forecast that the Indians would never accept the provisions of this law. Far from buckling under, the Indian prime minister has reacted to this threat of cutting off the supply by asking Indian scientists to press ahead with their experiments to produce a substitute for the enriched uranium from America. They then explained that they would spike fuel elements with locally extracted Indian plutonium. Here we have a concrete example of the point made in Giraud's statement.

The matter came to a head in May 1978. The U.S. Regulatory Commission had to decide whether U.S. exports to India could take place, and they decided by a tie vote of two to two, not to allow such an export. However, since Mr. Carter was to be visited by the prime minister of India a few days later, he decided to authorize the supply. Mr. Desai came to Washington, and I understand that the two states had a general "tour d'horizon" but that two main bones of contention were discussed. One was the question of the delivery of U.S. supplies to India, and the other was the question of Indian exports of rhesus monkeys. It seems that the Indians have a quasi-monopoly on the export of rhesus monkeys, which are used a great deal in labs all over the world for various experiments, including irradiation experiments, and I understand that the Indians, for religious reasons, have the strongest possible objection to the kind of treatment that is performed upon these unhappy animals. Desai came to Congress, which according to law could reverse the decision of the president and actually prevent the export of uranium and according to the press reports Mr. Desai did not budge one inch. Rather, he repeated to Congress that the reason India would not sign the nonproliferation treaty is that India feels that the treaty is discriminatory in favor of the super powers. When the big nuclear powers stop making nuclear weapons and begin to reduce their stockpiles, then India will consider signing a nonproliferation treaty. Desai did not indicate whether India would agree to put its two plutonium extraction plants and its two nuclear reactors under international safeguard, to make sure that plutonium would not be used for military purposes. He pointed out that the United States had a contract with India and that it is up to the United States to live up to its obligations. Finally, Congress gave way and did not reverse the executive decision of the president.

It seems to me that after this Indian episode and after the gentlemen's agreement between the U.S. authorities and the European authorities to renegotiate the U.S.–EURATOM agreement only insofar as the subject matter is not covered by the INFCE, this nonproliferation legislation has proved to be in its most negative effects a kind of paper tiger.

CONCLUSION

If we look at the history of proliferation, it is obvious that efforts to build atomic weapons have always been decided on political grounds that had nothing to do with the peaceful use of atomic energy. Therefore, if we are serious about the best ways and means to stop proliferation, we should look for political solutions to problems that are problems of security—military security and economic security. For instance, there is a point that has not been touched on at all so far that I think is very important—namely, the question of Japan. Here is a country that is highly advanced, has no energy, and still, in fact because of its own national history, does not seem at all tempted to go nuclear. And why is that so? Because by and large the Japanese people are protected by the U.S. umbrella. If they had the impression that this umbrella were going to be withdrawn or that there were leaks in the umbrella, then that would be the most dangerous temptation for Japan to go nuclear. In retrospect, you may look at certain U.S. decisions to withdraw from Korea as potentially very destabilizing insofar as Japan is concerned.

In conclusion, I believe that the more effective measures against proliferation are those that are trying to deal directly not with the effects but with the causes for a country's possibly going nuclear. For example, at the time I am speaking I understand that the president of the United States is meeting in Camp David with the leader of Israel and the leader of Egypt. If, as I hope, they arrive at some kind of solution—or, to switch to another continent, if one makes progress concerning the solution of the Rhodesian problem—then by such measures one will have made very important steps toward preventing Israel or South Africa—two states that many people think are in a position to explode a nuclear device whenever they wish—from going nuclear.

NOTE

1. Karl Kaiser, "The Great Nuclear Debate: German/American Disagreements," *Foreign Policy* 30 (Spring 1978): 83–113.

Chapter 7

Postwar Nuclear Relations: Lessons and General Applicability

*J. Robert Schaetzel**

The 1970s seem synonymous with change—the end of the postwar generation, the conclusion of twenty-five years of history that in turn ushers in a world of startling, even frightening instability. Not merely the pessimists but many of those who normally insist that things are not so bad as they seem stand deeply troubled by incipient international chaos, a sense that events, not man, are in charge. The world cries out for a degree of order.

It is instructive to look at the place and role of nuclear issues in contemporary affairs, not merely on their merits within a narrow framework, but as a laboratory case to test man's capacity to command events. The advantage of the nuclear issue, especially of those aspects related to nonproliferation, is that by being a discrete subject, it is intellectually more manageable. That nuclear proliferation is perceived by almost everyone as a clear and present—and universal—danger is a central factor. The explicit weapon potential of plutonium, the risk of diversion, and the general consensus about these dangers suggest that this is an area theoretically ripe for international agreement and for enforceable arrangements.

Since Hiroshima, a consensus has emerged regarding a series of

*Former U.S. Ambassador to the European Economic Community

factors that support the foregoing contention. First, only a minority, and a minority dismissed as insane, do not share the general fear of nuclear war, a war in which there would be no winners. Second, the problem seemed to be manageable by the fact that there were a limited number of nuclear powers. Third, the nature of nuclear weapons and the inherent impossibility of separating peaceful and military applications created a pattern of either extensive government ownership or at least extensive governmental control. Fourth, as a new field, technically novel, one could hope that the classical industrial vested interests had not yet taken root.

Fifth, and perhaps one of the most important factors, in an East-West atmosphere of mutual suspicion, hostility, and competition, this seemed to be one of the few fields in which the United States and the USSR had a clear common interest—controlling the spread of nuclear weapons. Sixth, these factors in turn produced a further asset—namely, that the universal apprehension about the danger of the materials and the threat of proliferation to world peace and stability meant that the public response would be to exalt the general good over narrow national interests.

Finally, there were certain institutional advantages. Drawing on American experience, an initiative was taken that led to the creation of the International Atomic Energy Agency, an agency whose assignment would be not only to assume responsibility for the controlled transfer of critical technology but also to a vehicle for managing special nuclear material. A parallel, regional institution came into being with the establishment of the European Atomic Energy Community (EURATOM). EURATOM provided the political and technical basis for collaboration between the United States and Western Europe in joint projects and the exchange of technology and material.

The foregoing is a rather imposing list of factors supportive of a unique and extensive international regime. But what assessment would we make at this time of these several factors? More than thirty years have passed since nuclear bombs were dropped on Japan. People have died, yet the disastrous genetic effects anticipated have not occurred. But most importantly, the memory of the horror of those weapons and their effects has faded. The fear of nuclear war is much less in evidence than the abandon with which governments and pundits advocate extending the infinite variety and, by implication, the attractiveness of arsenals of thousands of nuclear weapons. While there have been continuous efforts to organize public opinion against such weapons, these efforts have had little effect on general public opinion. Indeed, nuclear power and nuclear waste are the subjects that have generated the most articulate opposition, stimulated largely by environmental concerns. One can only conclude that the anticipated leverage of broad public concern about nuclear war did not develop.

The assumption that the traditional "vested commercial interests" would either not exist or be of slight consequence turned out to be entirely unrealistic. As power reactors entered the market place, governments found it necessary or inevitable to withdraw in favor of the private manufacturer. Furthermore, as economies faltered and the balance of payments became the preoccupation of governments, the governments themselves became allies of industry in the fight for commercial markets. Competition took another form, both among governments and among private enterprise, in support of alternative technologies by which nuclear power could be generated. In a curious but understandable way, this complex array of competitive factors created its own form of vested interests.

As far as the industrial democracies are concerned, the development, manufacture, selling, and maintenance of power plants are now largely in the hands of established multinational companies. That these by the nature of the business tend to be major concerns means that by definition they wield great power within their own governments. As a footnote, this is manifestly less than true today in the United States, where major companies such as General Electric and Westinghouse have been remarkably unsuccessful in exerting influence on either the Congress or the executive branch.

Events have also cast in doubt the belief that government ownership or at least extensive and special government control would make the international management of nuclear power easier. It also turned out to be a less than uniquely significant factor. The shift away from complete government control began early after the launching of the Atoms for Peace program in the United States. The process accelerated with the declassification of technology and has continued apace, particularly under Republican administrations. The end result is a nuclear industry in many respects indistinguishable from the normal commercial enterprise. One of the most notable changes from expectation to reality occurred in Europe, where the role envisioned for EURATOM began to erode from the very beginning and was superceded by the growth of national nuclear programs and national nuclear industries.

In retrospect, the fanciful assumption that the number of military nuclear powers would remain limited to only the United States and the Soviet Union seems now to have been wildly optimistic. The collaboration between America, France, and Britain in the development of nuclear weapons led to the United States' honoring its obligation to the British. The rebuff to the French probably accelerated their accession to the club of military nuclear powers. With India following China, the notion of a permanently small, limited fraternity of nuclear weapon states was hard to sustain.

Perhaps the hardest argument to overcome in pursuit of the goal of nuclear nonproliferation is this proposition: If the United States and the Soviet Union determined that nuclear weapons and the full fuel cycle were essential to their security and economic interests, why should other nations not arrive at exactly the same conclusion?

The assumption that the United States and the Soviet Union have special and almost peculiar common interests in maintaining their monopoly retains a certain validity. This has been a wary collaboration, colored by the immense suspicion that each holds for the other. In addition, each is tarred by the same brush—the charge of attempting to impose a double hegemony. In view of the way the world is organized and the fact that peace is a function of mutual terror, the charge of hegemony cannot be disregarded. Acceptance of the balance of terror is less a matter of worldwide appreciation of the "constructive role" of the two super powers than acquiescence, for the moment, with the status quo. The presumed parallelism is flawed by a fundamental asymmetry. The Soviet Union exercises effective control over its Eastern European satellites while the United States is at best an uneasy leader of a loose coalition, one member of which revels in its unpredictable independence.

The daring concept of international ownership and management was lost when the Acheson-Lillienthal initiatives failed. The decline in the capacity to generate large ideas was matched by the unwillingness of countries or peoples to embark on the kind of unique international enterprise needed to bring nuclear materials under effective control. By the 1950s the United States was unprepared to offer the unprecedented commitment of technology and material that as an effective monopolist it could have done, possibly with startling international effect. It is equally instructive to note the degree and the speed with which Western European nations turned their backs on the potentiality implicit in the EURATOM treaty to construct a truly supernational regime. And beyond that, they were quite unprepared individually or collectively to embark on any major innovative international action.

We have already seen how the presumed dominance of the public over the traditional private interests faded with astonishing speed. Today, while commercial concerns do not totally dominate the field, they have a major influence on national decisions. The decline in the installation of nuclear power facilities is more a function of a lower level of economic activity and the unexpected decline in the demand for electrical power than any explicit decision of governments or international institutions to slow down the process while a safer international regime is constructed. Another side of the same coin, as indicated earlier, is that with countries and concerns desperate to increase export earnings, a general willingness developed to bend the rules on security considerations for the sake of enhancing export opportunities. Again, the American case is an exception.

It turned out too that the developing countries were able to bring to bear special leverage on Western European countries and Japan. They indicated that their ability to purchase nuclear power plants and technology would be viewed as indicative of the general political and economic policies of these industrialized nations.

Until the twilight of the Eisenhower Atoms for Peace program, international excitement abounded about the benefits that nuclear energy would bring. The EURATOM Wiseman's report is only one of many examples of the translation of this enthusiasm into programs that turned out to be substantially romantic. The 1950s were an era of governmental and industrial salesmanship, a classic example of oversell. To a degree this unbridled enthusiasm contributed to the understandable, in political terms, reactive lack of enthusiasm of Eisenhower's successors. And then came the breakdown of the American nuclear establishment. The high quality and dedication of the Joint Congressional Committee for Atomic Energy and its uniquely competent staff began to decay. The Atomic Energy Commission lost its way in internal quarrels and debilitating battles with the Congress that finally resulted in the dismantling of the agency. This devolution was encouraged and has been accompanied by the efforts of well-organized and articulate antinuclear pressure groups.

As for American–European community relations in this field, as the momentum of the 1950s declined and the events discussed in the previous paragraphs unfolded, American interest in working with the European community slackened. For the Europeans, privileged access to American monopoly of technology and material became less important as the monopoly eroded. The assumption of mutual interest in a uniquely beneficial relationship dissipated as well. This arose in part from the failure of EURATOM to develop along the lines originally contemplated. These dreams were victims of the revival of European nationalism, generated by de Gaulle, which took the direct form of French attacks on the community, specifically against EURATOM. The other member states became happy participants in the institutional infanticide initiated by France.

Another factor of increasing importance in all aspects of American foreign affairs, but particularly in this field, is the determined independence of the American Congress. The Congress became less and less willing to go along with the policies of the executive branch. Congressional assertiveness has been notable with respect to nuclear energy.

U.S.–EURATOM collaboration came close to collapse with the Carter administration. Abrupt unilateral moves related to nonproliferation shook the foundation of U.S.–European cooperation and raised the gravest doubts about Carter's interest in nuclear power as such. In any event, foreigners ignore at their peril an American executive establishment and a Congress, each with a mind and will of its own, acting independently.

In sum, whatever may have been the opportunities and the momentum of the 1950s and early 1960s, they were lost for many reasons, including bad luck and changed circumstances. A decline in the quality and the imagination of both American and European leadership made it easier to miss the opportunities. When one recalls the consensus that existed across the Atlantic in the early 1950s and into the early 1960s, one can only be depressed by the failure to exploit the possibility of collective action toward enlightened nonproliferation policies.

Certainly one conclusion to be drawn from this recital throws into pessimistic perspective the likelihood that the nations of the world are apt to be successful in creating effective international institutions in other fields, where the conditions conducive to collective action are far less evident. This resume suggests that those dedicated to the goal of bringing nuclear energy under effective international control would do well to consider carefully the following propositions:

The first is to give due recognition to the forces of nationalism. Wartime and postwar collaboration created an illusion of its ebbing strength. A parallel error was the belief that the world had come to appreciate the need for international action. Nowhere is the collapse of these hopes more poignant than with respect to nonproliferation.

The interdependence of nations is inescapable in nuclear affairs. And even here, progress in collective international action has been modest. Certainly the suppliers of raw materials, technology, and finished products have been shortsighted. Frequently their behavior has been irresponsible; often their responses have been dictated not by reason or policy but by domestic pressures. They have been guilty of overestimating the leverage they thought monopoly had given them. One sees this in terms of Australia's erratic behavior in the export of natural uranium, with the Canadians acting with similar irresponsibility.

But the United States is the most culpable of the supplying nations in the Western world. We encouraged the development of nuclear power. We insisted that we would be a benevolent, reliable, and low cost supplier of fissionable material. Then our performance became at best quixotic. America as the unreliable supplier not only surprised and damaged allies, but undermined the chances of effective international action. And we perversely strengthened the arguments of those who were skeptical of collaboration with "foreigners," especially with the United States, and who were thus encouraged to seek security in autarky.

Misjudging the force of nationalism entailed specific costs. The supplying nations should have appreciated the unwillingness, and in certain cases even the political impossibility, of nonnuclear countries' accepting a nuclear regime that made discrimination blatant and subordination explicit. For example, to dramatize the distinction between the military nuclear and nonnuclear powers was bound to create extraordinary domestic political problems for any Italian government.

A review of this record makes it hard to see how we could have discounted to the extent we did the effect of the pervasive distrust existing between East and West. This distrust included such questions as the shadowy dominance of the Soviet Union over the Eastern European countries and the unwillingness of certain Western countries to accept the bona fides of Soviet or Eastern European IAEA inspectors. In such a difficult and sensitive area to have to cope as well with distrust is almost too much. Inevitably this suspicion has reduced the effectiveness of the agency and the willingness of other countries to convey to the IAEA greater responsibility.

Another factor, almost impossible to eliminate, flows from the asymmetry of Western and Eastern European political systems. The former is open, subject to continuing observation, examination, and criticism; the Soviet system is closed, cloudy, and conducive of Western distrust.

A further disturbing element is the matter of commercial secrecy. Secrecy for commercial advantage is a fact of economic behavior. If one thinks of proprietary interests as a spectrum, the United States is perhaps on the far end of a scale of relative openness, the Western Europeans somewhere in between, with the Soviet system effectively closed. Obviously, secrecy for commercial advantage inhibits any system of international control.

Perhaps one of the most impressive errors made in the early days was extravagant forecasts of the technical ease and relatively low cost of nuclear energy. While the dangers of nuclear power were appreciated in general from the beginning, it is interesting to contrast those halcyon days with the present suspicions and fears.

As the United States became more conscious of the security and safety requirements needed for the responsible development of this new source of energy, this concern was subject to easy misconstruction by competitors, who interpreted proposed international controls and restrictions as little more than thinly veiled attempts by America to secure commercial advantage. This has been particularly true in case of the breeder reactor and with regard to American efforts to delay the use of plutonium as a fuel. Other countries leaped to the conclusion that given the abundance of uranium available to the United States, its massive isotopic separation capacity, and the lack of a consensus in the United States on the economic and safety merits of the breeder, for the United States to oppose development of the latter system was an act of extraordinary American disingenuousness. Americans also misjudged the influence of private pressure groups on government policy, particularly in Europe, seen notably in the case of German arrangements with Brazil, but for a time equally evident in several putative French arrangements.

One lesson emerging from this record and generally relevant to national and international affairs is the need to manage and reconcile multiple goals, many of which are frequently in conflict one with the other. This emerges starkly in the field of nuclear energy—for example, in the face-to-face confrontation of nonproliferation policy with the goal of developing alternative energy sources that has become more and more urgent with the oil crisis.

The second lesson is that any effective control of nuclear energy requires a level of sophistication of international arrangements beyond any we have had in any other aspect of international life. The kinds of collaboration and self-denial required are in the sharpest conflict with the rise of nationalism and the related desire for national independence.

Third, around the world, and for quite understandable and commendable reasons, private groups determined to protect the human environment have appeared. These interest groups have directed much of their attention to the dangers of nuclear energy, particularly those of waste products. This brings them into opposition with those who urge the necessity for expanding nuclear power as an indispensable supplementary source of energy.

Fourth—and this gets into the heart of every international institution—is the intrinsic conflict between effectiveness and universality. With over 150 nations—a majority of which are presently or potentially involved in nuclear energy in one form or another—there is the immediate question of whether international institutions of universal membership have the slightest chance of performing efficiently and economically, if at all. On the other hand, if effectiveness demands a limited number, the organization becomes elitist and certainly nondemocratic.

A fifth factor is the inevitable battle between efficiency of the marketplace—in the production and sale of raw and processed fissionable materials and in the manufacture, sale, and management of industrial installations—and an intergovernmental system and intergovernmental management that under the best of circumstances will be of questionable efficiency and of certain high cost.

Sixth, we face the question of whether to attempt to devise a system of universal criteria uniformly applied or to work toward a common law regime of broad principles where the system results from the decision of one case after another. Seventh, should the system be composed of a series of regional arrangements, following the model of EURATOM or should it hold to the more traditional path of a pure international framework?

Eighth, there is the impact of "the have" versus "the have not" problem—in this case, nations which have the raw materials and technology versus the aspirations and resentment of the "have nots." This is part and parcel of the line that divides the world between the industrialized and the developing countries. It tends to be sharp and precise in this field. There is a special problem related to the interests of some of the oil-producing countries. Iran has been a striking example of this phenomenon, a country that saw that at some fixed point in the future its oil resources would be totally exhausted. In the interval it wants access to new technology and use of the revenues from this finite product to construct a base for new industries. These aspirations have brought several countries into direct conflict with nonproliferation policies.

Ninth, a subtle but most important point is the difficulty of establishing national strategies and priorities. When this phenomenon repeats itself in each of the major countries, it then becomes close to impossible to put together, out of this matrix of confusing and conflicting national policies, a coherent international system.

Finally, there is the difficulty of developing solid and agreed data and facts upon which an international consensus can be built. This is apparent in almost all aspects of the nuclear field—world reserves of uranium, the cost and feasibility of various methods of isotopic separation, the relative merits of the uranium versus plutonium fuel cycles, and so forth. And yet a firm consensus on these essential ingredients is indispensable to the development of an international strategy and an effective international organization.

This is a depressing story. The implications are discouraging, furthermore, for the eventual development of a truly effective and safe international economic and security system. But in all fairness, our postwar nuclear experience is not entirely discouraging. An effective and respected international agency has been developed and is in operation in this critical field. Unique among international agencies, the IAEA has developed a safeguard system that both the suppliers and the consumers respect. And an impressive degree of international consensus has been developed. In the West, the organization and the quiet work of the so-called "Suppliers Group" has made a unique contribution. To deal with the contentious problem of the uranium versus plutonium issue and the breeder reactor, the International Fuel Cycle Evaluation project was conceived. Not unimportantly, the United States has pulled back from some of the Carter blunders caused by an excess of innocence and misguided enthusiasm in 1977 and 1978. The American government accepted the fact, albeit with some reluctance, that if answers to many of the problems were to be found at all, this would occur only through a process of international collaboration, not in unilateral dicta.

It is important in this connection to identify the major role and the responsibility of the Europeans, and not only for their own sake, for seeing that a sensible and safe international regime is developed. It is essentially unfair as well as unwholesome to allow an international vacuum to occur that perforce only the United States is left to fill in what must be a search for some kind of sensible international nuclear order.

We are left with a basic question after this review of whether the international community can in fact learn from history and profit from its own mistakes. Certainly if we cannot develop an effective consensus and minimal system in this critical area, then one can only have the deepest pessimism about the likelihood that the international community has the elementary good sense and the limited will to solve other and even more complex world problems.

Chapter 8

Is Internationalization the Alternative to Nonproliferation?

*Russell W. Fox**

THE NATURE OF THE PROBLEM

The world has a problem, a very serious problem. It is a problem that derives from the fact that the same atom can produce both great destructive power and lasting injury and power for peaceful purposes. And it is the degree to which processes for both purposes proceed along the same lines that makes the situation acute. The existence and nature of the problem have long been recognized, but its intensity and reality have only recently received emphasis. As yet we have no answer to it.

Much—too much—has been written and said about the problem. Politics intrude at almost every point, but I wish to suggest that it can do with more objective analysis, and to that end would like to put forward certain matters:

ESSENTIALS OF THE PROBLEM

The first thing to say is that we have a nuclear industry and nuclear research activities. Countless billions have been invested in

*Ambassador at Large for Australia for Nuclear Nonproliferation and Safeguards

them, overall, in thirty or forty countries. We also have reprocessing and well-advanced plans for its expansion. Several countries have gone far in their steps to produce fast breeder reactors, and at this stage we have to assume that several of those reactors will be in operation before the turn of the century.

The next is that the risks are global: it is not simply the supplier countries that are at risk. The problem is not simply one of the actual spread of nuclear weapons; the fear that another country may use its capacity for military purposes is in practice of real concern. Fear begets reaction. What is needed is the machinery to create confidence.

A weapon can only be produced from fissile materials—plutonium, highly enriched uranium, or U-233 (which derives from the use of thorium). It is therefore fear of the risk of diversion of these substances for military purposes or of the misuse for military purposes of equipment capable of providing them that is of concern. Put shortly, the problem is the risk of diversion, and what we are mainly concerned with is the perceived risk. If it were possible to devise and operate a system in which spent fuel is kept, stored, and disposed of as such, the proliferation problem would be removed. This may be possible in some situations.

There is already a marked degree of international interdependence, perhaps a unique degree, so far as concerns the development of nuclear power industries. There are few if any countries that can, or at least that find it convenient, to develop and operate fuel cycles entirely from their own resources. This gives us an important guide to the nature of the solutions we must find.

A number of states see nuclear energy as essential for their survival or development, and it may be expected that more will come to do so. When, however, we talk of the need for nuclear power, it should not be regarded as the easy alternative. Conservation, which should include avoidance of waste, will be forced on us. When all these matters are properly considered, it is possible that a country will find that it can do without nuclear power or greatly limit its contemplated nuclear program.

Lead times are long, and planning must be well in advance—periods of up to twenty years are sometimes involved. Costs being enormous and planning problems considerable, it is most desirable that countries in need have adequate and timely assurances of supply. The time factor and lead times must also be borne in mind, in connection with the existing resources of countries, when assessing the risks of proliferation from any particular direction. Any manifestations of the problem will be largely in the future. The political tendency to limit considerations to the short term must be counteracted.

SOLUTIONS

The first thing to recognize is that there cannot be a solution in an absolute sense. A country that wishes to develop nuclear weapons, crude or otherwise, will do so sooner or later. But this is a serious political step that will become known to the rest of the world—which will react accordingly. Moreover, there may be some countries that will escape, or not satisfy, any nonproliferation regime that is established. In that case, the country can only be expected to be understood and treated on its merits.

Our main concern is to create confidence, the lack of which is so disruptive and can have such evil consequences. Assurances, and notice, are central to the question of confidence. The keynote can only be international cooperation; dictation by one or more countries is not an answer.

Every effort should be made, by scientific and technical means, to ensure that no ingredient of a fuel cycle is readily usable for weapon purposes. If coprocessing proves workable, it should be regarded as a normal part of the operation of any reprocessing plant of the future. If mixed oxide and plutonium fuels are used, they should, on present information, be irradiated. Other technical measures are already known and should be taken at the appropriate time.

The world has already gone a long way in effecting international arrangements—IAEA treaty and the organization supporting it, EURATOM, NPT, the Tlateloco treaty. It is plain, however, that these do not create the necessary international confidence. Recently, supplier states, recognizing their primary responsibility, have sought further ways to meet the risks. This has usually been done by means of bilateral agreements, the surveillance of which is usually left in the hands of the IAEA. The IAEA has produced two studies, one dealing with the advantages of regional (multinational) fuel cycle centers and the other with the control of plutonium stocks and spent fuel. One should not ignore, in this context, the study on physical protection. The International Consultative Group on Nuclear Energy (ICGNE), sponsored by the Rockefeller Foundation and Chatham House, has established an international study group on this issue.

It is undesirable for this matter to be left in the hands of supplier nations. The bilateral agreement system, whereby a purchaser may be required to submit to requirements imposed in the interests of nonproliferation by one or more of the countries involved in supplying materials or services at the front end of the fuel cycle, plainly has its problems. The position is, or may be, worse if a country is able to obtain its fuel without any such requirements. It may of course produce the lot for itself. And it is bad that supplier countries should be in competition among themselves so far as regards their safeguards provisions.

Countries must guard against the danger of trying to find one solution for all cases. Sufficient assurance of nondiversion can be obtained in many ways, although it will be necessary to bear in mind the need for sufficient international acceptance of what is done. Assurance of supply for a particular country can also be obtained in many ways, but this is more a matter for particular purchasers.

A central theme of this chapter is that once there is sufficient international acceptance that diversion is under control, the ordinary operation of the market will give nearly, if not all, of the assurance of supply that is necessary. The clog on supply is largely a product of the proliferation fear. And it is to be remembered that suppliers want to supply and to have assurance of their markets. A great deal of capital is tied up in the operations that go to make supply possible.

It is important that nuclear weapons states (NWS) recognize the particular advantages that they have because of the NPT and because most of the rest of the world is prepared without challenge to accept their status. In my view, those advantages should not be carried into their civil industries, which they should keep separate and bring under IAEA safeguards and otherwise treat from a safeguards point of view as they expect the civil industries or activities of nonnuclear weapons states (NNWS) to be treated. It is a question requiring further consideration whether, because they produce plutonium and have wastes, it will help or ultimately hinder nonproliferation if they acknowledge a responsibility to store spent fuel or to reprocess for NNWS.

But one should start at the beginning. The danger arises from plutonium (or U-233) or highly enriched uranium. These must be placed under international or multinational control. Depending upon the scope of the particular control, the sensitive operations should themselves be brought under international or multinational control. Diversion may still occur, but the control schemes will be an added deterrent, and in any event there will be immediate notice of the diversion. The controllers or coparticipants may be ejected, but this would be a serious step, of which the world would have prompt notice. If there is multinational participation, this can be left to be arranged and organized by the parties themselves. Adequate steps can be taken to ensure that trade and national secrets are adequately protected. The tendency should be to reduce the worldwide number of sensitive operations. Again, it is to be remembered that the particular arrangements may vary considerably; they must always have the same end in view, but should be adapted to meet the circumstances of the case. If help is needed, the IAEA should be available to advise and assist at any stage.

For this purpose, and perhaps for other purposes as well, it might establish an advisory committee or statutory body, appropriately constituted. The committee or body should in suitable cases have power to publish its advice.

There will still be questions about the assurance of supply. This is not a right that can be insisted upon without regard to the nonproliferation measures taken or given by the intended recipient. On the contrary, the price of assurance should be cooperation in nonproliferation measures. When this cooperation is forthcoming, it should be expected of potential suppliers that they give such assurances as they can and that are reasonable in the circumstances. In some cases the assurances can take, or be assisted by, concrete measures or by the erection of commodity banks, held on a trust or fiduciary basis. When it comes to the supply of uranium or enriched uranium, an intending user will often be able to secure greater assurance by becoming a shareholder or coparticipant in the enterprise in question.

As far as is practicable, non-proliferation standards should be standardized internationally, but reviewed periodically. The existence of the unusual or exceptional case is to be recognized. A country should, however, not require more than the proclaimed standards or offer less without the view of an IAEA committee having first been obtained and published.

I suggest that we begin by securing the internationalization of stocks of plutonium (whether coenriched or not) or of highly enriched uranium and prescribing them as steps to be taken at the appropriate time. There is nothing novel about such a proposal; many countries of the world had it in mind over twenty years ago (see Article XII A of the IAEA statute). The NWS should follow the example of the United Kingdom in undertaking that their civil stocks will be so treated. We can then see more clearly what should be done in each individual case. But nonproliferation and assurance of supply should be regarded by all parties as essential parts of the same activity. The disposal of spent fuel will doubtless come to be regarded as part of the same problem. Hence the need for international cooperation and coordination.

Chapter 9

Preventing Nuclear Proliferation: Scope and Limitations of International Action

*D. A. V. Fischer**

I propose to make some observations on the potential of and the inherent limitations to international action to prevent nuclear proliferation. The experience of the IAEA will serve as the model, but I shall also look at that of other international bodies such as the Nuclear Energy Agency of the OECD, CERN, and EURATOM, so far as I am conversant with them. Particular attention will be given to the IAEA's experience in nuclear safety matters and in applying safeguards and to the Non-Proliferation Treaty.

Perhaps some rather obvious conclusions can be stated at once. The first is that intergovernmental organizations like the IAEA are inherently well suited for setting safety standards and for certain related regulatory activities. International bodies do not have the power or the means to enforce standards and can only—and to a very limited extent—verify whether or not the international safety standards that they have set are being observed by enterprises within their member states. However, when it comes to verifying the conduct of governments (rather than entities within states, like the nuclear industry), international organizations can effectively verify observance of

*Assistant Director General for External Relations, International Atomic Energy Agency

agreed rules, provided that they have the means and necessary political support.

International organizations can play a useful role in promoting joint research undertakings—that is the pooling of resources from various countries into jointly owned and operated R&D or fundamental research projects. International organizations are inherently ill suited, however, to operate any kind of commercial enterprise. The closer that an R&D project comes to commercialization, the smaller are the prospects for effective international cooperation among more than two or three countries. As we have seen, even bilateral commercial technological ventures like Concorde and Airbus have their problems.

Let us look briefly at the experience of the IAEA against these generalizations. The statute authorizes the IAEA to undertake almost any conceivable action and to launch almost any undertaking to promote the peaceful uses of atomic energy. Within this broad framework, the statute foresees certain specific activities, namely:

1. Setting and applying safeguards against diversion of nuclear material to military purposes.
2. Setting standards for health and safety and applying them to assisted projects.
3. Serving as an international fuel authority for the pooling and supply of nuclear materials; and
4. Serving as a depository of plutonium to prevent national stockpiling.

Of these four types of activity, the two that are most fully articulated in the statute are the supply of nuclear materials, for which the IAEA was expected to serve as a pool or bank, and the application of IAEA safeguards. What has our experience been?

SETTING SAFETY STANDARDS

From the start in 1958, the agency has been reasonably successful in setting nuclear safety standards. In fact, this activity has grown considerably since the eruption of the so-called nuclear controversy. With regard to the enforcement or even verification of these standards, the agency's role has been much more modest. From time to time, there have been suggestions that the IAEA should be more active in ensuring that its safety standards are applied. The IAEA does make occasional checks by sending safety missions to member states or by arranging international evaluations of national projects, but the task of verifying the application of safety standards has not been developed to anything like the same extent as that of verifying that nuclear materials are not being used to further a military purpose. The reason for this can be summed up in one sentence: safeguards are applied to verify the conduct of governments; safety standards are applied by governments whose responsibility it is to verify the conduct of their own nuclear industry. The agency has neither the means nor the authority to take over responsibility for the safety and environmental protection of the citizens of any of its member states.

We must conclude that this will apply with equal, if not greater, force to the observance of measures of physical protection designed to prevent forcible seizure, high-jacking, and what is called subnational diversion. The IAEA can and is setting standards and is helping in the negotiation of a convention on physical protection. However, the enforcement of such standards must be the responsibility of the state concerned.

SUPPLY OF NUCLEAR FUEL—AN INTERNATIONAL FUEL AUTHORITY

The second major function foreseen for the agency—namely, that as a supplier of nuclear fuel—is quasi-commercial in character. It has never been more than a peripheral activity. Only in two cases have states turned to the agency for the supply of fuel for their nuclear power plants. In both instances, political motivations were probably strong—namely, a desire to interpose the IAEA between themselves and the major power that would actually provide the nuclear plant and material. When in one case the providing country decided to change the conditions of supply, the agency—although nominally the supplier—could do little more than serve as a channel to transmit the reactions of the recipient country. The resulting differences between the two countries were eventually resolved directly between them.

The agency's experience, therefore, is not conducive to much optimism about the value of internationally underwritten assurances of supply. Unless an international supplying authority has actual physical control of the material—a situation that is difficult to envisage in the real world—the value of such assurances will depend, in the final analysis, on the credibility of the supplying country, and this will, of course, be judged in part on previous experience. In the end, security of supply can probably be gained only if there is a wide diversity of suppliers in a fully competitive market.

MULTINATIONAL AND REGIONAL CENTERS

Pierre Huet, the first director general of the European Nuclear Energy Agency, drew some conclusions from his experience in the launching of joint undertakings. Three elements are crucial for success. The first is, of course, the political will of the participating countries. The second is a point to which I have already referred: The more remote the project is from a commercial undertaking—in other words, the more closely it is oriented to fundamental research—the better are its chances. The third conclusion was that the legal and political framework in which the project is set up must be flexible so as to permit it to change direction in the light of experience.

The IAEA's experience in launching regional or multinational projects involving hardware such as a laboratory or other facility has been limited and not very encouraging.[1] Since 1957, member states or the secretariat have come forward with about fifteen to twenty regional projects. In fact, even before the agency was established, the United States proposed in the early 1950s a $20 million regional research center in the Philippines that came to nothing. The first mission that IAEA sent out in 1958 was intended to launch a regional research project in Latin America. All the states visited agreed that it was an excellent idea, but with one proviso—that it be situated in their own territory. So far, we have scored only two somewhat modest successes out of some fifteen tries. In the early 1960s, a Regional Radioisotope Center for the Arab countries was launched under the auspices of the IAEA. In fact, this was the Egyptian National Center, which was being regionalized. The political will was mobilized by Egypt, which had a commanding position in the Arab League at that time. The second success was an international project for testing the wholesomeness of irradiated foods that we set up jointly with the OECD in the mid-1960s. This is still going strong on a modest scale in the Federal Republic of Germany.

The experience of other organizations is more extensive and pertinent. In the early days, the European Nuclear Energy Agency was successful in launching three regional R&D projects, two of which involved substantial investment. The first was the pilot reprocessing plant operated by Eurochemic, and the second was the Dragon experimental high temperature reactor in the United Kingdom. The Eurochemic project was effective in assisting its fifteen participating states to acquire the technology of reprocessing. Having served this purpose, there was an evident decline of interest in it, and it terminated three or four years ago. The Dragon project was very successful from a scientific point of view but came to what many considered to be an untimely end when the host government apparently lost interest in the technology of high temperature gas-cooled reactors.

All three projects were launched by NEA in the late 1950s and early 1960s. NEA also endeavored to launch several other projects, such as a regional heavy water production plant in Iceland and a European nuclear ship, but none of these bore fruit. As far as I am aware, the only new joint undertaking that NEA has fathered in the last fifteen years has been the modest Food Irradiation Centre that we set up jointly in the mid-1960s. NEA has since turned its energies very largely to problems of nuclear safety and waste management.

The only major international research project in (or rather, near) the nuclear field that has truly stood the test of time and is still going strong and expanding is that which is most remote from commercial application—namely, the experimental physics center at CERN with its accelerators.

I have not mentioned projects like COREDIF, EURODIF, and URENCO. The first two seem to me to be primarily national projects, although with large international capital participation, while URENCO is a commercial arrangement for pooling technological information and coordinating production and marketing in which there is no attempt to concentrate the hardware at one point. United Reprocessors is an even looser arrangement in this sense.

From this brief sketch, you are left to draw your own conclusions about the prospects of successfully launching large-scale multinational industrial undertakings for reprocessing or enrichment, such as the multinational fuel cycle center that has been under discussion for several years. You will gather that personally I am not optimistic. I feel, too, that the odds are similarly against any large-scale international commercial operation for the handling of spent fuel. If such projects are ever to see the light of day, they will have to follow the model of COREDIF or URENCO, rather than that of fully fledged, internationally constructed and operated undertakings like CERN. In other words, one would have to find a host government that was prepared to internationalize an existing national project (which it would build and operate anyway) or take the lion's share of responsibility for setting up a new one. I am not conversant with Eastern European experience, for which the ground rules are somewhat different, but note that the regional research center at Dubna is also devoted to fundamental research.

PLUTONIUM STORAGE

The fourth function foreseen for the agency—namely, to serve as a depository of separated plutonium—has remained quite dormant for the past twenty years. However, it has recently attracted a good deal of interest, and we have just circulated a study of the matter and will convene a meeting of experts from interested states late in 1978.

Given the necessary political will, it seems to me that international plutonium storage under the auspices of IAEA could be a viable and useful operation, provided that the quantities involved were relatively small and that the purpose was not to provide a commercial service. When large quantities of plutonium have to be dealt with, because of the commercialization of fast breeder reactors, I can see no alternative to a strictly commercial type of operation between reprocessors and utilities, subject, however, to strict international controls to prevent undesirable releases.

SAFEGUARDS

The function of the agency that has really "taken off" is, of course, that of applying internationally agreed safeguards in order to verify undertakings by states not to use nuclear material or plant to further military purposes. As we have seen, bilateral safeguards arrangements are at the mercy of abrupt changes in the political relations between the supplying and recipient countries. Although regional arrangements may be reasonably effective from a technical point of view, they do not have the international credibility needed to give the world community the assurances that it demands today. Regional safeguards arrangements that were once accepted as adequate by all the main Western industrial nations are no longer so regarded by several of these nations themselves. Thus, historically, it has been necessary to replace bilateral and to supplement regional safeguards by the international safeguards of the IAEA. There can be no going back.

We are, of course, aware of the limitations of international safeguards. They cannot forcibly prevent diversion. At best, they can provide an effective means for the prompt detection of diversion and, hence, a deterrent to a would-be diverter.

Much attention has been focused on the technical problems of safeguards, particularly in relation to reprocessing and enrichment plants, but experience is showing that they are all soluble. Moreover, as Walter Marshall once said, no government wants to be caught "with its hands in the till." If there is a fair chance that it will be caught red handed, it will seek a legitimate or at least nonclandestine way to achieve its objectives. Safeguards must ensure that there is a fair chance that the would-be diverter would be promptly caught. This objective is well within our technical reach.

The real problems of safeguards are political. They must be applied with equal stringency in countries that would not conceivably contemplate diverting and in those that have strong incentives to do so. They must command the financial and political support of the very diverse group of countries that run the international organization. At the operating level, inspected countries are able to cause considerable difficulties by obstructionists tactics of one kind or the other.

In short, the continuing success of international safeguards depends to a far greater degree on political support and, in particular, the political cooperation of the inspected countries than on whether or not the organization is able in all cases to reach its self-appointed technical objectives—for example, to detect within a prescribed period at a prescribed confidence level the diversion of a prescribed quantity of nuclear material.

The political value of safeguards to the international community is that they provide a continuing and credible assurance to other countries—adversary, neighbors, or friends—that the safeguarded country is not in fact diverting nuclear materials to military or other explosive uses. It is this assurance that constitutes the contribution of safeguards to international security.

I would like to stress this point, since a false proposition that has gained wide acceptance in academic circles in the last couple of years has found its way into national export policies and is even reflected to some extent in the U.S. 1978 Non-Proliferation Act. This proposition is that the chief political significance of safeguards is the length of warning time that they would give when a diversion has occurred and before the diverter has actually exploded his device. The "warning time" must be sufficient to give diplomacy "time to act" upon the would-be diverter. If the would-be diverter already possesses enriched uranium or separated plutonium either in a stockpile or in a reprocessing facility, the warning time may be too short. Hence, plutonium must be kept out of the hands of nonnuclear weapon states.

This "overnight plutonium grab syndrome" has distorted our thinking about fast breeder reactors, which really have very little to do with the backyard reprocessing plant that could be the real danger and have no connection at all with the small-scale enrichment plant that may be the route of the future to "nuclear weapons capability." This proposition also implies that there is little value in safeguarding those industrial countries that already have reprocessing (or enrichment) facilities or significant quantities of separated plutonium. In other words, there is little point in applying safeguards in most of the significant nonnuclear weapon states. Yet it is these very countries that undoubtedly have the capacity to make nuclear weapons.

In short, this new interpretation stands the whole concept of safeguards on its head, focusing on the risk of abrupt diversion in supposedly unstable Third World countries and deprecating the value of the assurance that safeguards provide about the nuclear activities of powerful technically advanced countries.

Let there be no doubt, however, that the explosion of a nuclear device by a seventh country would have serious consequences. Its impact on the Western nuclear power industry would be at least comparable to that of a major accident. The main attack on nuclear power today by critics like Lovins and Alfvén is that it inevitably leads to proliferation. This is demonstrably untrue. Nevertheless, the explosion of a nuclear device would be adduced as proof of their thesis. Since such an explosion would probably take place in a region of political tension, its politically destabilizing effects in the region would obviously be acute. Even more serious, however, might be the more remote consequences. Apart from the incentive that proliferation would give to adversary or competing countries to follow suit, it might become very difficult to hold the line among some of the major industrial nonweapon powers, particularly if the proliferating country were a relatively close neighbor.

The true takeoff point for IAEA safeguards was in March 1970, when the Non-Proliferation Treaty came into force. The early 1970s marked the peak point of international consensus on what to do about horizontal proliferation. The situation will never recur. The international nuclear industry will still very largely be dominated by one country in the West and, of course, by another in the East. When the two super powers reached agreement, the resulting force was difficult to resist. There was also still a near consensus on nuclear energy matters in North-South relations. The fruits of this were the treaty itself and the unanimous agreement subsequently reached between more than forty countries on the standard safeguards agreement to be used for NPT purposes.

Since then, the nuclear energy industry has become polycentric; the United States has practically dropped out of the market except as exporter of enriched uranium, and its place has been taken by Western European countries. The North-South dialogue, as it is euphemistically called, has become much sharper in tone, and the Indian explosion, the oil crisis, the nuclear environmental controversy, and far-reaching changes in the policies of what was the major exporter would make it impossible in the foreseeable future to achieve again a broad international consensus like that of the early 1970s.

If this is correct, it becomes all the more important to protect and strengthen what we have achieved and all the more hazardous to put this achievement at risk by abrupt changes in direction or by piling new institutional edifices upon it. We are already nearing a testing time. The second NPT Review Conference will take place in 1980, and the treaty itself will come up for renewal in 1995.

Let us look at the present situation. One hundred and four countries are now parties to the NPT or to the Tlatelolco Treaty or both. They include all the major industrial nonnuclear weapon states except one—namely, Spain. There are thirteen significant nonparties to the treaty—significant in the sense that each of them has at least a research reactor or other nuclear facility. They are Argentina, Brazil, Chile, Columbia, Egypt, India, Indonesia, Israel, the Peoples Republic of Korea, Pakistan, Turkey, South Africa, and Spain. All but two on this list are developing countries: several are in areas of political tension.

However, it is important to bear in mind that in eight of the thirteen, every significant nuclear facility of which we are aware is at present under IAEA safeguards because of the operation of non-NPT safeguards agreements. Thus, for instance, all significant plants of which we are aware in Argentina, Brazil, and Pakistan are at present under safeguards. These countries are, of course, not under any legal or treaty obligation that would prevent them from developing their own unsafeguarded facilities or fuel cycles, if they are able to do so.

What about the five other countries? The unsafeguarded plants in Egypt and Spain are not of strategic significance. Moreover, Spain will have to join the Non-Proliferation Treaty when it becomes a member of the Common Market. This leaves us with three countries that have unsafeguarded fuel cycles—India, Israel, and South Africa. Their nuclear status has been the subject of intense speculation, but one thing is clear: it is here that the danger of proliferation is most acute. In other words, the most pressing problem confronting us is not any deficiency in IAEA safeguards; rather, it is the absence of these safeguards in certain crucial areas.

The first objective must therefore be to extend the nonproliferation regime, de facto or de jure, to those countries having significant unsafeguarded facilities. The second objective relates to those eight countries where we already have de facto full-scope safeguards. Four of them are in Latin America, and if the NPT is not acceptable to them on grounds of principle, the Tlatclolco Treaty may offer an alternative. In fact, Argentina has already proposed exploratory talks in this connection. It does seem to me to be quite an achievement on the part of Latin America to have attained the status of being the only region in the world in which nuclear weapons or safeguards explosives are not being manufactured and that is also free from accusation of clandestine manufacture. It would be a vast achievement if this de facto status could be perpetuated and formalized by agreement between all the states concerned.

Obviously, it lies outside the power of the IAEA to achieve either the first or the second objective. The means are chiefly in the field of politics and nuclear commerce over which the agency has little control. We have frequently expressed the view, however, that no new agreement for the supply of nuclear material or equipment should be concluded with any nonnuclear weapon state unless it accepts full-scope safeguards through either the NPT or Tlatelolco or by some other legal arrangement with the IAEA.

I have referred to recent preoccupation with short "warning times" and the resulting distortion of the approach to the problems of plutonium, reprocessing, and fast breeder reactors. In part this results from the application of fanciful scenarios to unlikely situations. Of course, nobody would welcome the widespread diffusion of small reprocessing or enrichment plants, and we should do our best to slow down this process, though the task may not be easy. It can only be achieved by building on the legitimate nuclear industrial aspirations of countries like Brazil and Argentina and not by seeking to stifle them.

I have spoken about the probable consequences of any further horizontal proliferation and of the urgent need to obviate this risk by universalizing the nonproliferation regime. Unless we can avert the direct threat of proliferation from unsafeguarded fuel cycles, the value of the numerous ancillary measures that have been proposed is likely to be very limited. Almost all these projects—the Multinational Fuel Cycle Centre, the International Fuel Authority, the International Plutonium Management Project, guarantees of supply, as well as the technological fixes that have been put forward such as Civex, coprocessing, and nonproliferating enrichment processes—are essentially intended to remove temptation from those that have already declared themselves to be virtuous by joining the NPT or Tlatelolco. This is, of course, not without value. But as the Bible tells us, it is even more important to redeem those who have already sinned or may perhaps be on the point of sinning.

NOTE

1. On the other hand, the International Centre for Theoretical Physics, which the agency established in Trieste in the early 1960s and in which all member states take part, has been an unqualified success in promoting cooperative research in this field—which is as remote as can be from practical applications and involves little hardware beyond a blackboard and chalk. The research center on the effects of radioactivity in the sea, which the agency established in Monaco in the late 1950s, has also played a useful role as an "umpire" laboratory, providing, in effect, standards and carrying out standardized procedures to check the performance of national laboratories.

Eberhard Meller is Principal Administrator in the Office of Long-Term Co-operation and Policy Analysis, Alternative Energy Sources Division, International Energy Agency of the Organisation for Economic Co-operation and Development in Paris. Prior to joining the OECD, he was Deputy Head of the Energy Supply Industry Division of the Federal Ministry of Economics in Bonn. Educated at the Universities of Heidelberg, Berlin, and Geneva, he received a Ph.D. in comparative law from the University of Heidelberg and Lyon. Dr. Meller is the author of many published articles about international energy affairs.